FUSARIUM SPECIES:

An Illustrated Manual for Identification

FUSARIUM SPECIES

An Illustrated Manual for Identification

Paul E. Nelson,
T. A. Toussoun,
and
W. F. O. Marasas

The Pennsylvania State University Press
University Park and London

Preparation and publication of this book was supported in part by Contract No. 223-78-2018 from the Department of Health and Human Services, Public Health Service, Food and Drug Administration, Washington, D.C.

Figures 40c, 42c, 44a–d, 45a,b,e, and 73a were originally published in L. W. Burgess, P. E. Nelson, and T. A. Toussoun, "Characterization, geographic distribution, and ecology of *Fusarium crookwellense* sp. nov.," *Transactions of the British Mycological Society* 79(1982): 497–505; they are used here by permission of the publisher.

Library of Congress Cataloging in Publication Data

Nelson, Paul E., 1927–
Fusarium species.

Includes bibliography and index.
1. Fusarium—Identification. 2. Fungi—Identification.
I. Toussoun, T. A., 1925– II. Marasas, W. F. O. III. Title.
QK625.T8N44 1983 589.2'4 82-62197
ISBN 0-271-00349-9

Designed by Dolly Carr
Printed in the United States of America

Contents

Preface

The members of the genus *Fusarium* are among the most important plant pathogens in the world. In recent years the genus has acquired additional importance as many *Fusarium* species have been shown to produce mycotoxins causing both animal and human diseases. One of the problems encountered by workers interested in *Fusarium* species, and particularly in toxigenic *Fusarium* species, is the correct identification of each strain. This problem was the impetus for the preparation of this volume. A companion volume, *Toxigenic* Fusarium *Species: Identity and Mycotoxicology,* serves as a catalog for information on *Fusarium* species already reported to be toxigenic in published reports.

The taxonomy of the genus *Fusarium* has been a subject of controversy for many years. We have taken the 1935 publication, *Die Fusarien,* by Wollenweber and Reinking, as a starting point. Since 1935 several other taxonomic systems have been proposed for the genus. These systems vary from one with more than ninety species to one with nine species, with several other systems in between these two extremes. All of these systems are based on the work of Wollenweber, and although each system has something to offer, none of them is satisfactory by itself for the identification of all *Fusarium* species. In this volume we have selected the best features of several systems and combined them in a compromise system which will allow the individual worker to identify *Fusarium* species using several of the current taxonomic systems. We have examined sufficient material and cultures to present the reader with detailed descriptions and illustrations of 30 species. We provide less-detailed descriptions and illustrations of an additional 16 questionable species because we have not been able to obtain and examine sufficient material to determine their authenticity. The reader should not assume that these are the only *Fusarium* species. Rather, they should be viewed as a starting point. Additional species will undoubtedly be added by other workers and over the years this system will be subject to change. We hope students, researchers, and specialists will find this volume useful.

We thank Nancy Fisher, Lois Klotz, and Laurie Morelli of the Fusarium Research Center for their invaluable assistance which made this project possible.

Our special thanks to to Prof. Dr. W. Gerlach, Biologische Bundesanstalt fur Land- und Forstwirtschaft, Institut fur Mikrobiologie, Berlin-Dahlem, West Germany, who generously provided us with cultures of many of Wollenweber's species, as well as some recently described by himself and his co-workers. In addition he supplied us with a complete set of the drawings made by Wollenweber and published as "Fusaria autographice delineata," and allowed one of us (TAT) to work in his laboratory with the wealth of material collected and stored there. We are most grateful for his help and material which provided us with an insight to Fusarium taxonomy that could not have been obtained from anyone else.

We thank our colleague Dr. L. W. Burgess, Department of Plant Pathology and Agricultural Entomology, University of Sydney, Sydney, Australia, whose wide-ranging surveys of crops and soils have provided us with many of the interesting and unusual cultures used in this project.

Additional cultures have been supplied to us by many other colleagues, among whom are C. Booth, R. J. Cook, and L. D. Dwinell. We express our sincere appreciation to all who have helped in this way.

The contribution of the South African Medical Research Council in granting a 6-month leave of absence to Dr. W. F. O. Marasas to assist in the preparation of this book is gratefully acknowledged.

We are grateful to Mr. R. E. Ackley of Photographic Services, The Pennsylvania State University, who took the color photographs.

Finally, we express our appreciation to the director of The Pennsylvania State University Press, Chris W. Kentera, and his fine staff. This volume marks our third effort with the Press and the association has been a most rewarding one for the authors. The consideration, encouragement, and assistance of the Press staff always makes our task easier.

FUSARIUM SPECIES:

An Illustrated Manual for Identification

Introduction

This book is a practical guide to the identification of *Fusarium* species according to the taxonomic systems of H. W. Wollenweber and O. A. Reinking (74, 76), W. Gerlach (21), A. Z. Joffe (35), C. Booth (4), W. C. Snyder and H. N. Hansen (44, 58, 59, 60, 65, 68), and C. M. Messiaon and R. Cassini (41). A complete discussion of all of the taxonomic systems for *Fusarium* species is given in Part IV of this book. Persons interested in this genus should be able to identify their specimens to species according to the taxonomic system of their choice regardless of their prior expertise.

In order to accomplish this the directions for growing cultures given in this book must be followed. These methods and techniques are based on 30 years of concentrated effort on the problems of growing *Fusarium* species for identification. The reader is forewarned that deviation from the procedures outlined will render identification of specimens more difficult or impossible and will decrease the usefulness of the information in this book.

Our basis and starting point for Fusarium taxonomy is the monumental work of H. W. Wollenweber (74) and his colleagues culminating in Wollenweber and Reinking's *Die Fusarien* (76), published in 1935. Accordingly we have organized this book by using their sections and species in the same order as they are presented in *Die Fusarien,* except in the sections Sporotrichiella and Liseola. In these sections species are arranged on the basis of recently discovered morphological characters. The treatment of more recently described species is based on the work of W. Gerlach (21), who continues the Wollenweber tradition in the same laboratory in Berlin.

While we have based our identification for the most part on the species of Wollenweber and Reinking (76), we have not recognized their varieties and forms, but have included them within the primary species. Species in the sections Elegans, Martiella, and Ventricosum are based on the species as emended by Snyder and Hansen (58, 59). Species in the section Liseola follow Wollenweber and Reinking (76) but are emended where necessary according to Gerlach (21) following Nirenberg (47). These modifications are based on our own experimental work and knowledge of variability in this genus. We have tried to be practical in our approach and have made our observations with the standard light microscope, using the media and techniques detailed in this book. Wherever possible we have worked with large numbers of isolates (i.e., more than 50 specimens/species) freshly gathered from a wide geographical area. These isolates and the mutants which arose in the laboratory in the course of subculturing were examined. The features that they had in common formed the basis of our judgment as to species characteristics and species limits. Species so treated form the bulk of this book and are included in the synoptic keys.

Not all of the species treated by Wollenweber and Reinking (76) are included. Many, particularly tropical forms in the sections Eupionnotes, Macroconia, Submicrocera, and Pseudomicrocera, have not been seen since their first description. These, together with certain species described by Booth (4) also not seen by us, are not included. In addition a few of Wollenweber and Reinking's (76) species, together with a number recognized by Gerlach (21), are omitted from the synoptic keys and are placed at the end of our taxonomic treatment because of the paucity of material available to us for examination and/or the degenerated condition of the cultures we examined. What remains, however, are 30 species representing the bulk of the more commonly encountered *Fusarium* species.

Synoptic keys to the sections and species as well as detailed descriptions and illustrations of 30 species of *Fusarium* are presented. We have been able to obtain sufficient material of each of these species to describe them with confidence. The species names are those used by Wollenweber and Reinking (76) unless otherwise indicated. Following the species name, synonyms and the taxonomic systems they represent are listed. We have covered the taxonomic systems of Booth (4), Gerlach (21), Joffe (35), Messiaen and Cassini (41), Snyder and Hansen (44, 58, 59, 60, 65, 68), and Wollenweber and Reinking (76) in this section. The species names in each of these systems are also listed in the index. Thus the description of any isolate can be located either by using the synoptic keys, if the culture is not identified, or the index, if one has the species name as given in any of the six taxonomic systems covered in this book.

Part I
Techniques for Isolating and Growing *Fusarium* Species

Isolation of *Fusarium* Species from Different Substrates

Fusarium species are a widespread cosmopolitan group of fungi and commonly colonize aerial and subterranean plant parts, either as primary or secondary invaders. Some *Fusarium* species are common in soil and it is rare to find a necrotic root of a plant in most agricultural soils that is not colonized by at least one *Fusarium* species. In many cases *Fusarium* species are wrongly assumed to be the cause of a disease because of their frequent isolation from necrotic roots, crowns, and stems. Many *Fusarium* species have the ability to colonize tissue as primary or secondary invaders, depending on the substrate. The majority of the species are fast-growing on culture media, but there are a number of important pathogens which are relatively slow-growing and thus can be difficult to isolate. Many of the isolation techniques described below were developed at the Fusarium Research Center, The Pennsylvania State University (68) and the Fusarium Laboratory, University of Sydney, Australia (6).

Isolation from plants

According to Burgess et al. (6), the choice of isolation procedures depends on the nature of the plant samples, the number of samples, and the *Fusarium* species involved. Isolation may be accomplished from a single sample or a few samples collected at one time, or it may involve many samples collected during systematic surveys for a particular disease over several growing seasons. The most common method for isolating *Fusarium* species from plant material is by culturing colonized material on agar media. However, there are a number of other techniques that may be used.

If the pathogen is producing sporodochia on the surface of the plant tissue, the sporodochia can be removed and used to prepare a spore suspension in sterile water. The *Fusarium* species in question may then be isolated using the single-spore technique for obtaining pure cultures.

If fertile perithecia are being produced on the plant tissue, a small piece of tissue with perithecia is well washed, the excess water removed, and the piece of tissue placed on the inner side of an inverted petri dish containing water agar or carnation-leaf agar; the inverted petri dish is then incubated under conditions of high humidity. The tissue piece may be held in place with vaseline or similar material. After incubation for about 24 hr, ascospores are released and land on the agar surface where they germinate and produce colonies. The fungus can then be transferred to other media by a single-spore transfer or by means of a hyphal tip.

Fusarium species, especially slow-growing species, may be difficult to isolate directly from necrotic root tissue which is extensively colonized by other fungi and bacteria. However, these species can sometimes be isolated using a combination baiting/plating technique. After the necrotic tissue is washed thoroughly and cut into small pieces, the pieces are mixed with steam/air-treated soil. Seed of the plant involved is planted in this mixture and incubated under conditions favorable for disease development. The plants are sampled as soon as root lesions develop and the root lesion cultured on suitable media after thorough washing or, if appropriate, surface disinfestation. This procedure has been used successfully to isolate *F. avenaceum* from old necrotic roots of subterranean clover (9).

The isolation of *Fusarium* species from diseased plant tissue on agar media is one of the most common methods used. The tissue selected should be typical of the diseased material under study. Areas of tissue that show extensive advanced necrosis should be avoided because this tissue is more likely to be colonized by secondary organisms. Tissue samples should be treated with a surface disinfestant to reduce the numbers of secondary organisms that may grow out on the agar medium and reduce the chances of isolating the primary organism. Sodium hypochlorite (Clorox) is a suitable chemical disinfestant to use on most tissue. Concentrations of sodium hypochlorite used range from 1 to 5% and the time of treatment varies with the nature of the tissue. Treated tissue should be damp-dried on absorbent paper toweling before being placed on agar to reduce the amount of bacterial contamination. Fine roots or feeder roots that are too small to be surface disinfested may be washed in running tap water for several hours and then rinsed in several changes of sterile water, damp-dried, and placed on the agar medium.

Culture Media Used in Isolation

Natural media. Isolation of *Fusarium* species from plant tissue should be done on a medium low in nutrients. Natural media such as those recommended by Hansen and Synder (31) and Synder and Hansen (62) are excellent for this purpose. They used plant materials such as wheat straw, barley straw, pea straw, and seeds embedded in water agar. Carnation-leaf agar may also be used.

Selective media. Nash and Snyder (43) devised an excellent medium for the direct isolation of *Fusarium* species from plant tissue and freshly collected field soils. The formula for modified Nash-Snyder medium is given below.

> 15.0 g Difco Peptone
> 1.0 g KH_2PO_4
> 0.5 g $MgSO_4 \cdot 7H_2O$
> 20.0 g Agar
> 1.0 g Pentachloronitrobenzene (Terraclor)
> 1.0 liter water

The medium is adjusted to pH 5.5–6.5, autoclaved, and 20 ml of a streptomycin sulfate stock solution and 12 ml of a neomycin sulfate stock solution are added to each liter after the medium has cooled and just prior to pouring into petri dishes. The stock solutions are prepared by adding 5 g of streptomycin sulfate to 100 ml of water and 1 g of neomycin sulfate to 100 ml of water. It is best to allow the medium to dry in the petri dishes for 5 days before use.

This medium is useful in isolating *Fusarium* species from plant material, especially material that is badly rotted or infested with fast-growing contaminants. As fungus colonies develop on this medium, large amounts of ammonia are given off which may kill the fungi if they are left on it for more than 20 to 30 days.

Another selective medium that is useful is a modified Czapek-Dox medium (6, 66). The formula for this medium is given below.

> 20.0 g dextrose
> 0.5 g KH_2PO_4
> 2.0 g $NaNO_3$
> 0.5 g $MgSO_4 \cdot 7H_2O$
> 1.0 g yeast extract ("Vegemite" or "marmite" may be substituted)
> 1.0 ml 1% ferrous sulfate solution
> 20.0 g agar
> 1.0 liter water

The following components are added after autoclaving and cooling to 48 C.

> 5.0 ml 1% suspension (0.05 g/liter) of 50% w/w 2,6-dichloro-4-nitroaniline (Botran)
> 0.1 g streptomycin sulfate
> 0.01 g aureomycin sulfate

The antibiotics are added from stock solutions to each liter after the medium has cooled and just prior to pouring and the medium should be stored in a cool dark place. Colonies of *Fusarium* species show more color on this medium than on the Nash-Snyder medium.

Isolation from Soil

Fusarium species may be isolated directly from soil either by using the dilution plate method or by culturing organic debris on selective media; they may be isolated indirectly from soil by using a root-baiting technique or sterile baits such as pieces of cereal straw (6).

Soil dilution plate technique. The soil sample is ground in a mortar and diluted 1:1,000 to 1:10,000 in 0.1% water agar. A 1 ml sample of this soil suspension is uniformly dispersed over the surface of a selective medium in a petri dish. Propagules in the soil suspension usually germinate in a few days and produce small colonies in a week. The dilution plates should be incubated in the light to insure sporulation. It is important that soil be air-dried before the suspension is prepared to reduce bacterial contamination. The agar medium should also be allowed to dry for 5 to 7 days prior to use for the same reason.

Both selective media mentioned earlier can be used for direct isolation from soil suspensions. The Nash-Snyder medium (43) suppresses most contaminants but does not provide colonies with vivid color. The modified Czapek-Dox medium (6, 66) provides more color in the developing colonies but does not suppress contaminants as well as the Nash-Snyder medium. Komada's (36) medium is useful for isolating *F. oxysporum* but will suppress the growth of some other *Fusariun* species. The formula for the basal medium is given below.

> 1.0 g K_2HPO_4
> 0.5 g KCl
> 0.5 g $MgSO_4 \cdot 7H_2O$
> 0.01 g Fe-Na-EDTA
> 2.0 g L-Asparagine
> 20.0 g D-Galactose
> 1.0 liter Water

When the basal medium is melted and cooled, the following antimicrobial supplement is added and mixed thoroughly.

> 1.0 g Pentachloronitrobenzene (PCNB) (75% W.P.)
> 0.5 g Oxgall
> 1.0 g $Na_2B_4O_7 \cdot 10H_2O$
> 0.3 g Streptomycin sulfate

The pH is adjusted to 3.8 ± 0.2 with an approximately 10% solution of phosphoric acid.
Debris isolation technique. A technique has been developed for the isolation of *Fusarium* species from small pieces of debris from soils (6). The soil sample is washed through a nest of three sieves, respectively 2.8 mm, 0.85 mm, and 0.5 mm in aperture. The first sieve retains identifiable plant remains, such as roots and crowns which can be surface disinfested and cultured on a selective medium. The small pieces of debris retained on the other two sieves are also cultured on a selective medium. However, the small pieces of debris are porous and cannot be surface sterilized easily. To remove as many contaminants as possible the sieves with retained debris are placed under a fine spray of filtered tap water for 2 hr or until the soil adhering to the debris has been removed. The debris is allowed to dry on sterile paper toweling before culturing. The debris can also be air dried before culturing to further minimize bacterial contaminants. Colonies of *Fusarium* species develop in 5 to 7 days and a wide spectrum of *Fusarium* species can be isolated with this technique. Both the Nash-Snyder medium (43) and the modified Czapek-Dox medium (6, 66) can be used.

Isolation from Air

The most widely used method is the exposed petri dish technique. Petri dishes containing a selective medium are exposed for 10 to 60 minutes, depending on spore concentration and wind conditions (33, 46). The petri dishes are then incubated under standard conditions and colonies are transferred as they develop.

Other workers have used spore traps to study *Fusarium* species in the air. Ooka and Kommedahl (49) used an Andersen sampler with petri dishes containing a selective medium to trap spores of *F. moniliforme.* Lukezic and Kaiser (37) used a Hirst spore trap to detect macroconidia of *F. semitectum.*

Growing *Fusarium* Species for Identification

The species in the genus *Fusarium* are highly variable because of their genetic make-up and because changes in the environment in which they grow cause morphological changes. Since morphology, especially the morphology of the macroconidia, is the basis for identification, it is necessary to take all possible steps to standardize procedures to make the task of identification easier. *Fusarium* species can be identified with certainty only if the cultures have been grown under optimum conditions for sporulation. All cultures used should be initated from single conidia, except in special cases which will be discussed later. Since each conidium is of a single genotype, the colony that develops, barring mutations, is of that genotype. It is a clone recognizable by its own particular characteristics. Mutant cultures which arise can be recognized readily. In this manner clones can be maintained indefinitely in culture.

While to the uninitiated all Fusarium macroconidia look alike, the real problem is that to the taxonomist they may all look different and a given isolate may seem to have macroconidia of several species. The most imperative problem, then, is to reduce this phenotypic variation as much as possible. This can be done by following the procedures outlined in this book; if these procedures are followed carefully the full value of this book will be realized and the photographs will be directly comparable to the investigator's preparations.

Culturing Methods

Single-spore method. The single-spore technique, devised by H. N. Hansen and modified by others (68), consists of pouring 3 ml of 2% water agar into unscratched petri dishes and allowing the agar to solidify. A suspension of conidia is prepared in a 10 ml sterile water blank so that it contains 1–10 conidia/low-power (10X) microscope field when a drop from a 3 mm diameter loop is examined on a slide. Experience enables one to gauge this concentration by simply glancing at the turbidity of the suspension in the water blank under the dissecting microscope. This suspension of conidia is poured over the solidified agar so as to cover the entire surface, and the excess is drained off. The dishes, thus seeded, are incubated in an inclined position at room temperature for 16 to 24 hr; they are then opened, shaken to remove the accumulated moisture, and examined under a dissecting microscope. Small squares of the agar containing single germinating conidia are cut out with a dissecting needle having a flattened tip and transferred to the desired growth medium. If the original culture is contaminated with bacteria, a drop of 25% lactic acid may be added to the 10 ml sterile water blank to inhibit bacterial growth. The acid spore suspension should be allowed to stand for 10 minutes prior to pouring on an agar plate. Use of this technique may delay germination of Fusarium conidia for 24 hr or more. Alternatively, solutions of antibiotics may be incorporated in water agar. The fungus is then allowed to grow up through the agar in a petri plate.

Hyphal tip method. Mutant colonies sometimes develop from single conidia taken from sporodochial cultures. It may be possible to produce new sporodochial cultures from such parents by hyphal tipping from the original colony. By this we do not mean a mass transfer from the growing edge of a colony, but the transfer of a single hyphal tip, under the dissecting microscope, in much the same manner as transferring a single germinating conidium. Mycelium from the sporodochial parent is used to initiate a colony on a medium low in nutrients, such as water agar. The agar is poured very thinly so that a sparse thallus develops. Hyphal tips can then be removed without difficulty. Clones which are known to cause problems through mutation should always be regenerated from colonies grown on weak media such as carnation-leaf agar or water agar.

Culture Media

Potato dextrose agar (PDA). When made according to our specifications, PDA is a valuable medium principally for gross morphological appearances and colony colorations. *Fusarium* species blossom out in their full diversity and color on this medium. Indeed, at times, they can be so striking as to be misleading. Cultures that superficially look alike can be different species, and cultures that look different may be of the same species. Particularly striking cases of such confusion can arise in the case of the carmine red color so often produced by certain *Fusarium* species on PDA. This striking coloration should be relied on with great caution for it is not a clear-cut criterion of species delimitation. For example, *F. equiseti* and *F. acuminatum* cannot reliably be differentiated on the basis of this color since certain clones of *F. equiseti,* called *F. scirpi* var. *compactum* by Wollenweber and Reinking (76), produce the red pigment. For the same reason it is not the factor separating *F. longipes* from Wollenweber and Reinking's *F. scirpi* var. *filiferum,* nor does it help in differentiating *F. lateritium* from Wollenweber and Reinking's *F. stilboides* (76). The problem also exists in *F. sambucinum*. Because of its high available carbohydrate content, PDA generally emphasizes growth to the detriment of sporulation. Cultures grown on PDA sporulate poorly, frequently taking more than a month to do so, and the conidia produced are often misshapen and atypical. Consequently, with very few exceptions (e.g., the microconidia of species in the section Sporotrichiella), PDA cultures are not used for microscopical observations. Cultures grown on PDA are thus used only in a secondary role.

Mutations are also enhanced when *Fusarium* species are grown on PDA and similar media. Transfers made from such cultures will gradually or suddenly sector and mutate. This is the principal reason why laboratory cultures degenerate in time. This type of variation is the bane of Fusarium identification. It alters culture descriptions, types of spores produced, and in extreme cases even the production and morphology of the macroconidia. The problem is minimized by subculturing as little as possible, by using the single-spore or hyphal tip techniques, and by not subculturing or storing fungi on media high in carbohydrates such as PDA.

This medium should always be prepared from raw ingredients rather than using any of the several commercial preparations available. Potato dextrose agar is prepared by using baking grade, white-skinned potatoes; red-skinned potatoes should not be used. The potatoes are washed and sliced, unpeeled, and 250 g of potatoes are added to 500 ml of water and placed in the autoclave along with a flask containing 20 g of agar in 500 ml of water. The potatoes are cooked and the agar melted by operating the autoclave on steam bypass for 45 min at 8 lb pressure. The potato broth is strained through several layers of cheesecloth into the flask containing melted agar. The remaining potato pulp is squeezed through several layers of cheesecloth until ½ cup of potato pulp is obtained; this pulp is then added to the melted agar and potato broth along with 20 g of dextrose. If necessary the total amount is brought up to one liter by adding water, and all the ingredients are then mixed thoroughly. The medium can then be dispensed into test tubes, autoclaved, and slanted. Properly made, each tube should have a small button of sediment at its base.

Carnation-leaf agar (CLA). We have based our identification procedures almost exclusively on CLA for the reasons given in the previous paragraphs. Almost all of the microscopic features are based on cultures grown on CLA, and all of the black-and-white photographs in this book are of structures formed on CLA. The great advantage of CLA is that it promotes sporulation rather than mycelial growth. Conidia and conidiophores are produced in abundance, their morphology closely approximates that seen under natural conditions, and phenotypic variation is reduced. All of the macroconidia in a given microscope field, for example, will be quite similar, as is evidenced by many of the photomicrographs in this book. This uniformity makes the work of identification simpler and avoids many difficulties. The value of CLA as a growth medium for *Fusarium* may be in the fact that it is low in available carbohydrates, and that it contains complex, naturally occurring substances of the type encountered by *Fusarium* in nature; consequently the fungi grow and sporulate in a manner similar to that found on a host plant or natural substrate. Indeed, carnation leaves, as we prepare them, are a natural substrate.

Carnation leaves may seem to many to be a rather esoteric substrate but we have found them to be eminently suitable for growth and sporulation of *Fusarium*. Therefore, we urge all those interested in identification to use them, so that the photomicrographs in this book will be directly comparable to the investigator's specimens. Where absolutely necessary other plant materials can be substituted, provided they are sterilized and handled in the same manner. While we have not had a great deal of experience in this, other green herbaceous material may be found to be suitable. We have found that leaves and stems of corn, wheat, alfalfa, and various other grasses can be used. Seeds are not recommended because they are generally too high in available carbohydrates and are more difficult to sterilize. We also strongly urge investigators to be consistent in their use of their chosen substrate as each different plant material has a modifying effect on the shape of the conidia.

Young carnation (*Dianthus caryophyllus* L.) leaves are harvested from actively growing, disbudded plants free from pesticide residues. The leaves are cut into pieces approximately 5 mm^2 and dried in an oven at 45 to 55 C for 2 hr. When dry the leaves are green and crisp. Loss of green pigmentation indicates that the drying temperature was too high. The leaf pieces are placed in aluminum cannisters 5 cm deep and 9 cm in diameter and sterilized with 2.5 megarads of gamma irradiation from a Cobalt 60 source. Propylene oxide fumigation may be used as an alternate method of sterilization (31, 68), but sterilization of the green leaves is not as thorough as with gamma irradiation and repeated fumigation may be required.

Carnation-leaf agar (13) is prepared by placing several sterile leaf pieces in a petri dish or culture tube and floating them on 1.5 to 2% water agar cooled to 45 C. The medium is left at room temperature for 3 to 4 days before use to check for the growth of possible contaminants from the leaf pieces.

Direct microscopic observation of a *Fusarium* species growing on carnation-leaf agar also indicates the manner in which conidia are borne on conidiophores. Use of this medium also favors development of perithecia of homothallic *Fusarium* species (69). **KCl medium.** The observation of the formation of microconidia in chains in the section Liseola is facilitated on KCl medium (14). This medium is prepared by adding 4 to 8 g of KCl to a liter of 1.5% water agar. Those species that form chains of microconidia form more abundant, longer chains on this medium. In addition the chains are easier to observe because there is less moisture on the surface of the agar and fewer droplets of moisture in the aerial mycelium. Direct observation under a microscope of 4- to 5-day-old cultures in petri dishes will demonstrate whether or not chains of microconidia are formed, and may show the presence of monophialides or polyphialides.

Formation of Conidia

Fusarium species produce macroconidia in sporodochia and in the aerial mycelium. In species in which they occur, microconidia are produced in the aerial mycelium. Macroconidia in sporodochia are produced on monophialides. Depending on the species, macroconidia and microconidia in the aerial mycelium are produced on monophialides or on polyphialides. While macroconidia produced in sporodochia are highly diagnostic, the observation of the structures borne in the aerial mycelium is also vital to species determination.

Macroconidia. The central characteristic feature of most *Fusarium* species is the shape of the macroconidia produced in sporodochia. This was recognized by Wollenweber and is the foundation of the Wollenweber and Reinking (76) taxonomic system. It is also the basis of this book. Sporodochia of most *Fusarium* species are produced rapidly and abundantly on CLA and the macroconidia so produced are uniform in shape. Cultures reach their optimum in this respect in 10–14 days. Identification always starts with such preparations. Macroconidia produced in the aerial mycelium are more diverse and thus more confusing. Nevertheless they should be examined also, particularly since certain species may not produce sporodochia. However, wherever possible, emphasis should be placed on macroconidia produced in sporodochia.

Microconidia. These are produced in the aerial mycelium, but never in sporodochia. Microconidia differ from the macroconidia in size and shape. While there inevitably is a certain amount of overlap, the typical sickle-shaped conidium is a macroconidium. Microconidia for the most part are 0–3 septate and are generally smaller. Difficulties arise in species having 3–4 septate, spindle-shaped conidia, but these are few and fairly easy to recognize. The presence or absence of microconidia, whether in chains or in false heads, borne on monophialides or polyphialides are important diagnostic features. The presence or absence of chains in species of the section Liseola are best seen on the KCl medium. The different types of microconidia in certain species of the section Sporotrichiella are sometimes better seen on PDA.

Conidiogenous cells. In *Fusarium* the conidiogenous cell in a sporodochium is a monophialide. In the aerial mycelium, conidiogenous cells giving rise to macroconidia or to microconidia may be monophialides or a mixture of monophialides and polyphialides, depending on the species. While the presence of polyphialides is an extremely important diagnostic feature, we do not recognize the presence of polyblastic conidiogenous cells in *Fusarium* as proposed by Booth (4). We feel that the demarcation is not clear and that there are no practical methods of differentiating the two. Provisionally, therefore, we recognize only monophialides and polyphialides in *Fusarium*. Polyphialides are found only in the aerial mycelium. Generally it is best to look for them when the cultures are 4–7 days old. Spore formation in the aerial mycelium falls off rapidly in older cultures and the conidiogenous cells lose their cytoplasm and are more difficult to see under the transmitted light microscope. They can still be clearly seen under phase contrast. With species in the section Arthrosporiella, polyphialides are best seen when cultures are 10–14 days old. It is important to remember that species that produce polyphialides also produce monophialides. It is also important to note that the characteristic of a polyphialide is a cell with more than one pore. Frequently, but not always, these pores are surrounded by a collarette which makes them more visible. It is important to look carefully for these pores, and such observations can be difficult. If only one opening is seen, the structure is a monophialide, not a polyphialide. A septum observed between two pores indicates branching but not a polyphialide. Sometimes *Fusarium* species will form monophialides which branch immediately beyond the pore before the septum is formed. When such a branching has just begun the structure may have the outward appearance of a polyphialide and can be quite deceiving (Fig. 20c–e). The presence of polyphialides is a dependable characteristic, in this respect second only to the shape of the macroconidium. However, it too is susceptible to the vagaries of mutation. In mycelial mutants with little sporulation, polyphialides may be sparse and difficult to find; in pionnotal mutants the reduction in the amount of aerial mycelium can lead to similar difficulties. By and large, however, under normal conditions and with cultures freshly isolated or not subject to the mutational pressures of a medium with high levels of available carbohydrate and frequent transfers, polyphialides, when present, are formed abundantly and are easy to find (Fig. 16a).

Conditions for Growth and Sporulation

Sporulation and pigmentation are favored by light, including ultraviolet wavelengths, and by fluctuating temperature conditions (62, 77). If possible, all cultures should be incubated in an alternating temperature of 25 C day/20 C night. Cultures are incubated in diffuse daylight from a north window or in light from fluorescent tubes. We have found that F4055/SX fluorescent tubes (Consumer Lighting Products, Inc., P. O. Box 5760, Baltimore, Maryland 21208) are excellent substitutes for natural light. Two of these tubes in an ordinary 40-watt fluorescent fixture are suspended 40–45 cm above the laboratory bench or shelf supporting the cultures. A day length of 12 hr is sufficient. Cultures that do not sporulate readily may be placed under a fixture containing one 40-watt black light tube and one of the 40-watt fluorescent tubes described above to enhance sporulation. This light combination also enhances the formation of the perfect state (teleomorph) in culture. A useful mobile light bank for growing cultures has been described by Burgess et al. (6).

While fluctuating temperatures are best, *Fusarium* species grow well at constant temperatures of 21 to 22 C; the major exception is *F. nivale,* which does best at fluctuating temperatures of 12 to 18 C.

Chlamydospore production. Chlamydospores are generally produced in culture. However, a method for their more rapid production is to place a small piece of the fungus, together with some of the PDA medium on which it is growing, in sterile distilled water. Chlamydospores will form, in some species that produce them, in about 7 days. Alexander et al. (1) have also used sterile soil extracts, obtained from poor sandy soils; it is prepared by mixing 1 liter of water with 1 kg of unsterilized soil, filtering through glass wool, and sterilizing by passing through a 0.22 µm millipore filter. Soils higher in organic matter may also be used if the procedure outlined in their paper is followed.

Observation and photomicrography. Since stains and fixatives frequently distort spores, it is preferable to use water as a mounting medium for examining conidia. A less temporary preparation can be made with a 0.15% solution of gelatin. This is an excellent method for the preparation of mounts to be photomicrographed; indeed, all the photomicrographs in this book were made by this method. Circular cover glasses are used, and the mounts can be kept for 15 to 20 minutes. As the mount gradually dries the cover glass presses down so that all the spores are in one plane and often line up through surface tension. A layer of gelatin accumulates at the periphery of the cover glass, slowing down the evaporation of water. Observations should be made rapidly because the macroconidia tend to swell due to water imbibition. These mounts can be kept for longer periods in moist chambers. The spores will germinate, particularly at the edges of the coverslip, and the process can be followed under the microscope.

The photographs reproduced here were taken on a Leitz Ortholux Research Microscope with a Leitz 40X apochromat objective and a 10X eyepiece. The camera was a Leitz 4 × 5 Aristophot giving a final magnification of 1000X on Kodak Contrast Process Panchromatic Film. The negatives were contact-printed on Kodak Azo paper.

Cultural variation. The majority of *Fusarium* species isolated from nature produce their macroconidia on sporodochia. The sporodochial type often mutates in culture and in nature. These mutants in turn may give rise to others, so that a mutational sequence is developed. In pathogenic isolates these mutants frequently exhibit a loss in virulence, and loss of toxin production may also occur. Variability and its effect on virulence and taxonomy have been discussed in detail by several researchers (34, 50, 51, 56, 57, 63, 72). The mutation sequence has never been experimentally shown to reverse itself. Starting from the sporodochial type, mutation in general proceeds in two opposite directions: i) toward forms producing abundant aerial mycelium but few macroconidia, termed mycelial types (Plate 6b; cf. sporodochial type, Plate 2c), and ii) toward forms producing little or no aerial mycelium but abundant macroconidia, termed pionnotal types (Plates 1d, 8d; cf. sporodochial types, Plates 1c, 2c). In the mycelial type there is frequently a lack of sclerotia, sporodochia, and pigmentation, so that these mutants often have a white, featureless look. Macroconidia are sparse and as a result identification is rendered more difficult. The macroconidia of the pionnotal types, on the other hand, are formed from unbranched monophialides in sheets over the surface of the colony. The colony has a shiny, wet appearance. These cultures are often more highly colored than the sporodochial type from which they arose. The macroconidia can be longer and thinner or shorter than those of the parental types.

Basically the avoidance of mutants involves i) single-sporing of cultures, ii) hyphal tipping, iii) avoidance of media rich in available carbohydrates, and iv) keeping subculturing to a minimum. Long-term storage is best in liquid nitrogen or by lyophilization.

Control of culture mites. These can be excluded from culture tubes by means of cigarette paper barriers which are fixed to the mouth of the tube with a simple gelatin glue which contains $CuSO_4$ as a microbial growth inhibitor (61).

Colors Produced on Potato Dextrose Agar

Colors used in the synoptic keys and the descriptions of species are illustrated in Plates 1–8. A photograph of the upper surface and undersurface of the slant of each culture is shown. These plates illustrate shades of the colors that are used to describe the aerial mycelium, sporodochia, sclerotia, colony surface, and the color in the agar when viewed from the undersurface. The reader should remember that it is beyond the scope of this book to show the entire range of colors found in cultures of *Fusarium* species and that no single culture can show the typical colony color for an entire species; the color plates are selected to provide a reference point in determining the color of a particular culture. The reader is further reminded that these colors are representative only of cultures grown on PDA under the conditions described above; any alteration in medium or environment will introduce the possibility that colony colors will be very different from those described and illustrated in this book. The color names and the plates that illustrate them are listed below:

white—Plates 3d; 4c; 6b
cream—Plates 5c,d; 8a
orange—Plates 5a,b
tan—Plates 4a–d; 5a–c; 6a; 7d; 8c
brown—Plates 4a–d
reddish brown—Plates 2a–d; 8b,d

carmine red—Plates 1a–d; 2a–d; 3a–d
pink—Plates 3a–d; 6d
purple—Plates 6c; 7a–c
blue—Plates 6d; 8b
blue green—Plate 8c

Preservation and Storage of Fusarium Species

A method for long-term preservation of cultures of *Fusarium* species by means of lyophilization has been developed at the Fusarium Research Center, The Pennsylvania State University (13). Isolates to be lyophilized are grown on carnation-leaf agar in petri dishes for 7 to 10 days, and then checked for adequate growth and lack of bacterial contamination. Bacterial contamination is assayed by observation of a slide mount of each culture under the microscope, or by growth in tryptic soy broth (Difco Laboratories, Detroit, MI 48201). Each lyophilization run is prepared under sterile conditions in a transfer chamber. Several colonized carnation-leaf pieces are transferred to each of five replicate sterile 5 ml vials labeled with the isolate number. A 0.5 ml aliquot of sterile skim milk (Difco

Laboratories, Detroit, MI 48201) is added to each vial. The vials are loosely stoppered with split rubber stoppers, which allow for evacuation of air. The stoppered vials are placed in a tray and quickly frozen by pouring liquid nitrogen into the tray. A lucite plate slightly larger than the tray is placed on top of the partially stoppered vials. A VirTis drying chamber (Model No. 10-MR-SA, The VirTis Co., Gardiner, NY 12525) on a refrigerated freeze-dryer is used for lyophilization. The tray is placed on the pre-cooled (−35 C) shelf in the drying chamber. After 10 minutes, vacuum is pulled in the chamber and maintained at a reading of 10 µm Hg on a McLeod gauge. Shelf refrigeration is then turned off and the shelf heat is turned on to 15 C for 16 to 20 hr, while the samples dry gradually.

After lyophilization, the vials are sealed under vacuum by inflation of a rubber diaphragm in the chamber over the tray, which presses down on the lucite plate and forces the rubber stoppers to seal the vials.

After lyophilization, vials are capped and labeled showing the isolate number and date of lyophilization, and the vials are stored at −30 C. Recommended storage of lyophilized cultures is 5 C or lower (32) and some fungus cultures are stored at temperatures as low as −52 C (42). The lyophilized pellet from one replicate vial is cultured immediately on water agar or carnation-leaf agar to check for viability and possible contamination.

Storage in liquid nitrogen of cultures grown on carnation leaves is another recommended procedure particularly for isolates that sporulate poorly and for *F. nivale*.

Procedure for Identification

There are synoptic keys to the sections and to the species. In addition to the keys the descriptions and the photographs offer further explicit aid. Thus the investigator has several avenues of approach and there is nothing wrong with coming out of the keys with two or three possibilities to be resolved with the help of the photographs and species descriptions. The synoptic keys can be entered at any point, with as many characters as the specimen exhibits. It is important to remember that the keys are based on averages. As an aid to getting these averages, particularly of conidium shape, it is helpful to use the lower power of the compound microscope (ca. 100X total magnification) and to take a quick view of the field. In this case first impressions are most useful and avoid the trap of seeing too many variations.

Plate 1: a, *Fusarium decemcellulare;* b, *F. lateritium;* c, *F. longipes;* d, *F. longipes* (pionnotal mutant).

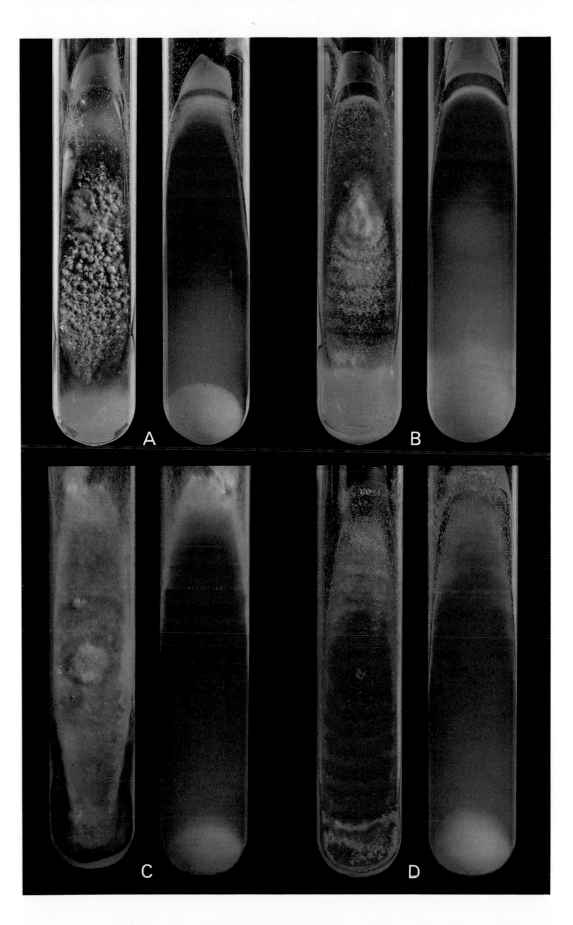

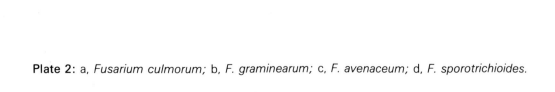

Plate 2: a, *Fusarium culmorum;* b, *F. graminearum;* c, *F. avenaceum;* d, *F. sporotrichioides.*

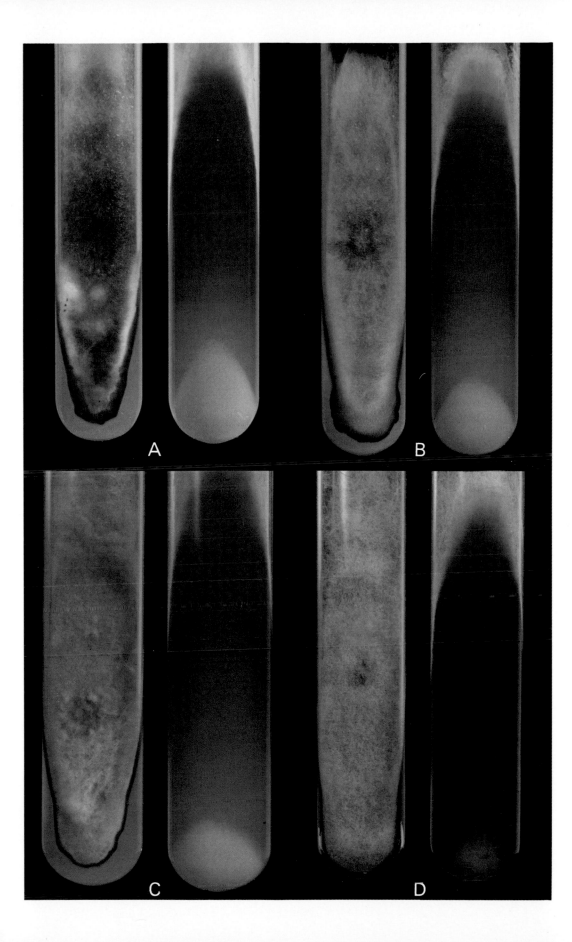

Plate 3: a, *Fusarium sambucinum;* b, *F. acuminatum;* c, *F. reticulatum;* d, *F. poae.*

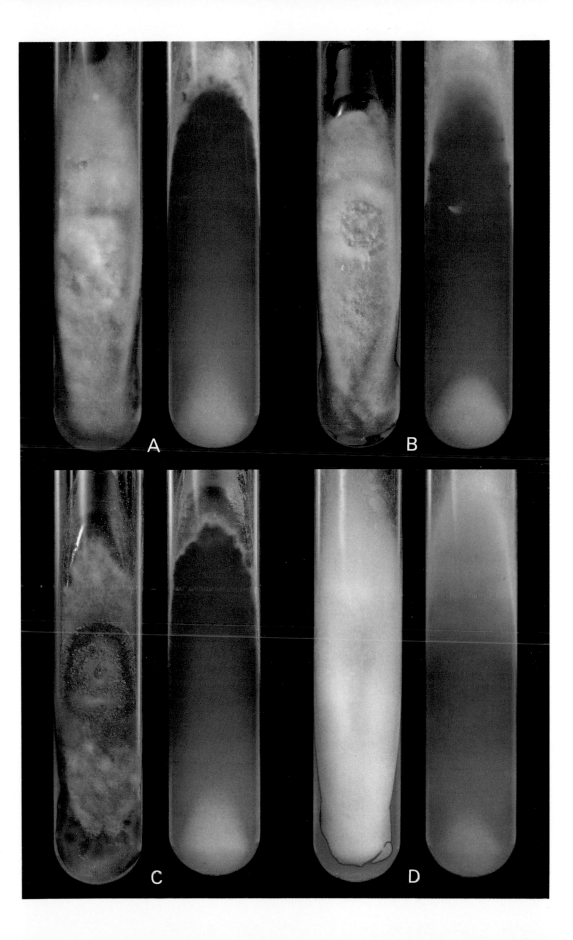

Plate 4: a, *Fusarium semitectum;* b, *F. equiseti;* c, *F. equiseti* (filiferum type); d, *F. chlamydosporum.*

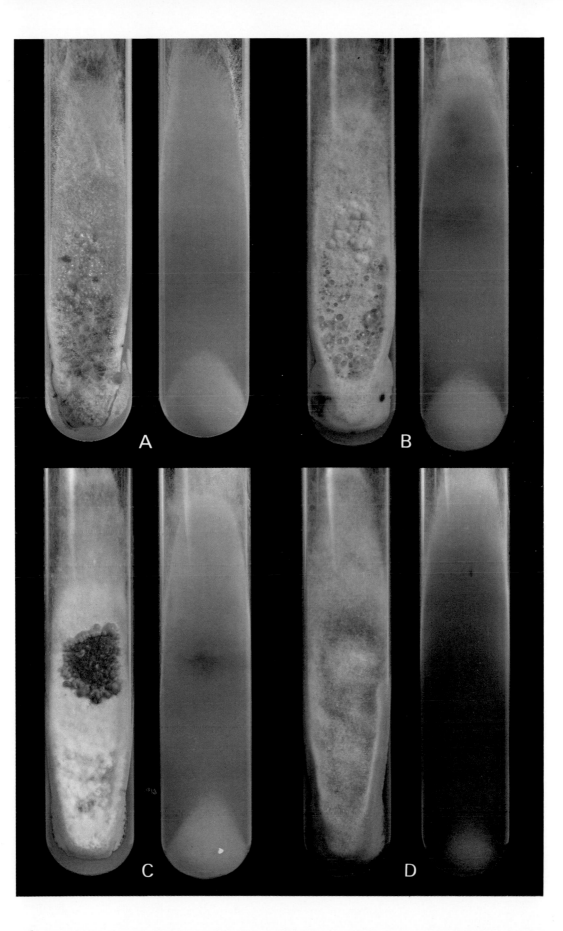

Plate 5: a, *Fusarium sambucinum;* b, *F. graminum;* c, *F. dimerum;* d, *Fusarium* species.

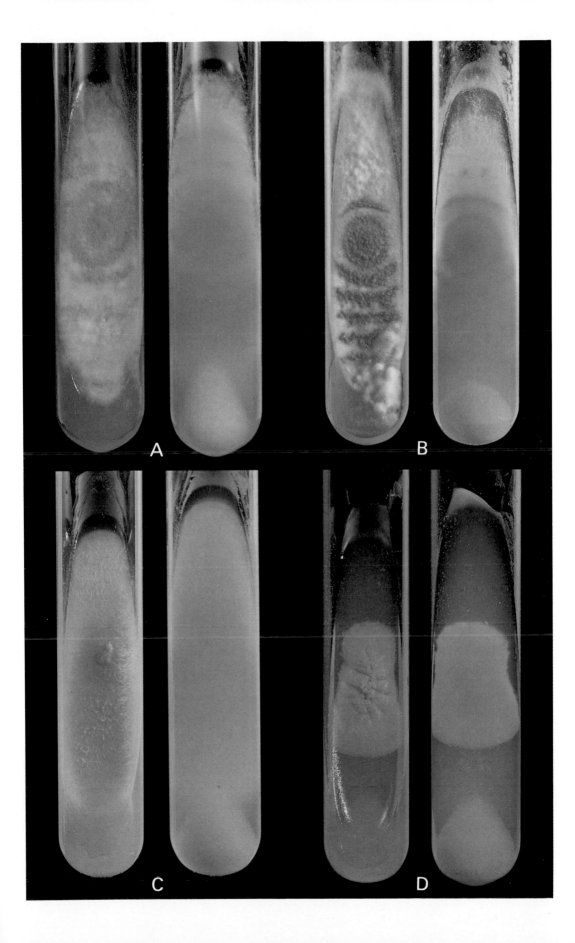

Plate 6: a, *Fusarium nivale;* b, *F. avenaceum* (mycelial mutant); c, *F. oxysporum;* d, *F. lateritium.*

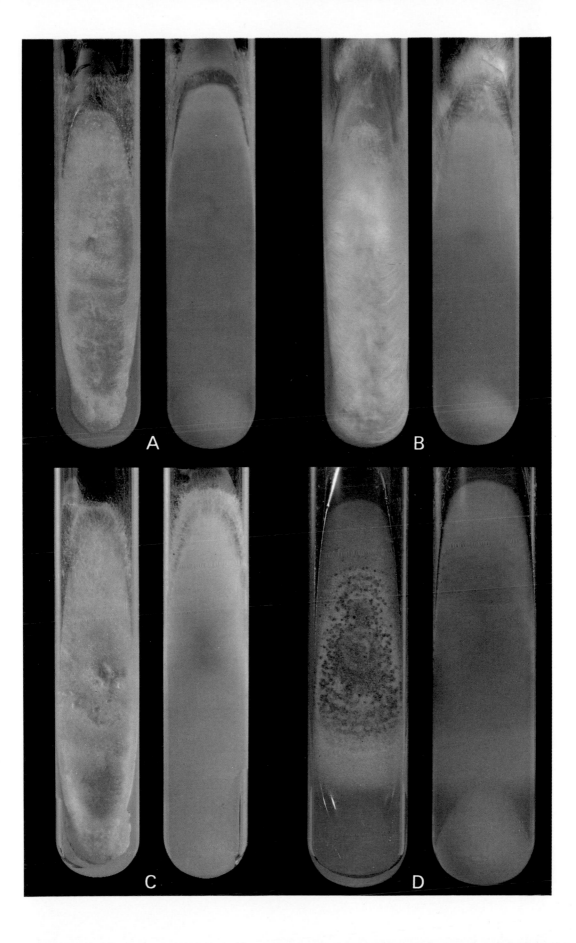

Plate 7: a, *Fusarium subglutinans;* b, *F. moniliforme;* c,d, *F. oxysporum.*

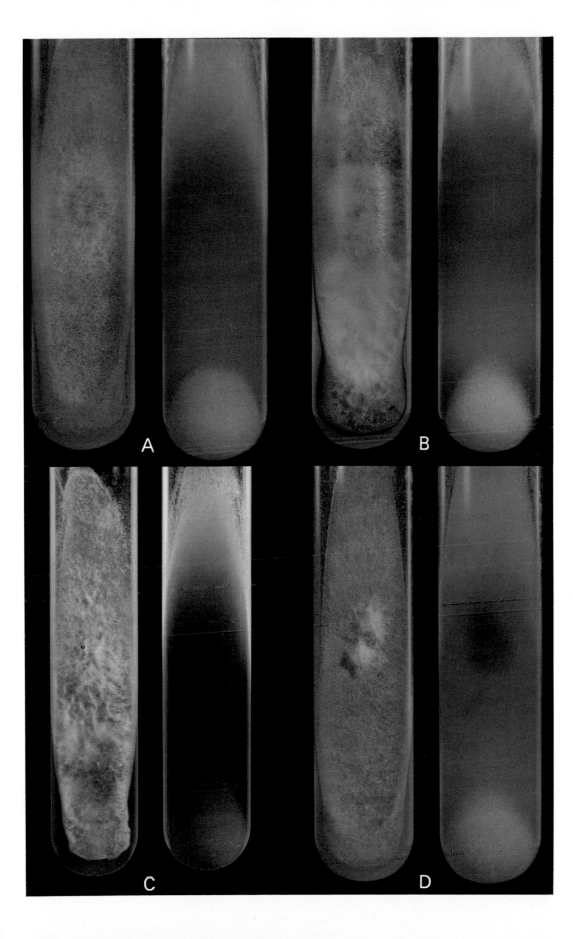

Plate 8: a,b,c, *Fusarium solani;* d, *F. avenaceum* (pionnotal mutant).

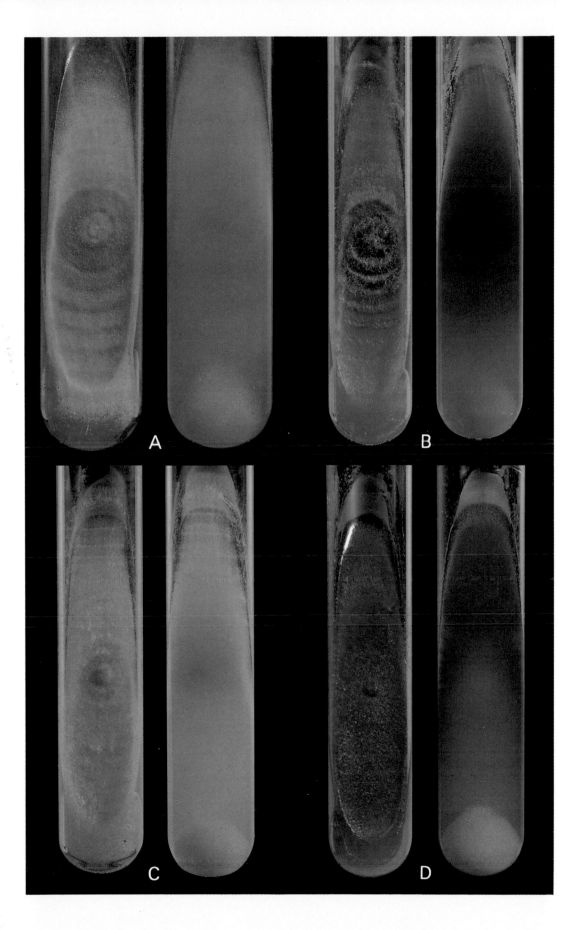

Part II
Synoptic Keys

Synoptic Key to the Sections

Sections are arranged in the same order as in *Die Fusarien* by Wollenweber and Reinking (76), except that sections Macroconia, Submicrocera, and Pseudomicrocera are omitted.

1. Eupionnotes
2. Spicarioides
3. Arachnites
4. Sporotrichiella
5. Roseum
6. Arthrosporiella
7. Gibbosum
8. Discolor
9. Lateritium
10. Liseola
11. Elegans
12. Martiella Ventricosum

As an aid to the reader, the numbers used to identify each section in the synoptic key appear in front of the section name in the descriptions of species in Part III. Numbers printed in italics indicate sections included in more than one description under a given characteristic. This key, of necessity, deals with generalities of form and color and does not cover all of the variations possible. It attempts to cover the more general features and should be useful for identification of at least 90% of the specimens.

Cultural characteristics are based on 10- to 14-day-old cultures grown on PDA slants. The descriptions of conidia, conidiophores, and chlamydospores are based on 10- to 14-day-old cultures grown on CLA plates. All cultures were grown from single conidia from sporodochial isolates and, when possible, from isolates recently obtained from substrates in nature. Mycelial and pionnotal mutants are not included in the descriptions. The illustrations of species should be used for direct comparison of conidium and conidiophore morphology when attempting to identify mutants. The presence or absence of microconidia and their shape should be checked on both CLA and PDA, especially in species in the section Sporotrichiella. Chlamydospores may form slowly in culture, often requiring one month or more, and both the CLA and the PDA preparations should be checked for the presence of chlamydospores.

SECTIONS

A. Cultural characteristics

1. Rate of Growth
   a. Growth very slow, cultures less than 2 to 3 cm in diameter in 10 days. *1*
   b. Growth moderately slow, cultures not more than 7 cm in diameter in 10 days. *2, 3, 8* (*F. reticulatum*), *9*
   c. Growth rapid covering entire (8.5 cm) length of slant in 10 days. *2, 4, 5, 6, 7, 8, 10, 11, 12*

2. Aerial mycelium present or absent
   a. Aerial mycelium present, sparse to felt-like or abundant, spore masses (sporodochia) present or absent. *2, 3, 4, 5, 6, 7, 8, 9, 10, 11, 12*
   b. Aerial mycelium absent, and the surface of the colony has a slimy, yeast-like appearance. *1*

3. Color of aerial mycelium
   a. White. *2, 3, 4, 5, 6, 7, 8, 9, 10, 11, 12*
   b. Light purple. *10, 11*
   c. Tan. *4, 7, 8*

4. Color of colony from below. Color may diffuse into agar.
   a. Color absent or if present only very pale shades of orange, tan, brown, or light purple. *1, 3, 4, 5* (*F. graminum*), *6, 7, 8, 10, 11, 12*
   b. Shades of carmine red. *2, 4, 5, 6, 7, 8, 9*
   c. Strong purple pigment diffusing into agar often in advance of colony. *10, 11, 12*

5. Color of spore masses (sporodochia)
   a. Cream. *11, 12*
   b. Orange to yellow to tan. *1, 2, 3, 4, 5, 6, 7, 8, 9, 10, 11*
   c. Reddish-brown. *4, 5, 8*
   d. Blue-green to blue. *12*

B. Macroconidia from sporodochia

1. Size
   a. Short, generally 1–2 septate. *1* (occasionally), *3*
   b. Medium long, generally 3–7 septate. *1, 4, 5, 6, 7, 8, 9, 10, 11, 12*
   c. Very long, generally 8–9 septate. *2*

2. Shape
   a. Macroconidia with marked dorsi-ventral curvature. The sides of the macroconidia are often unequally curved. *1, 3, 4, 5, 6, 7, 8, 11*
   b. Macroconidia without marked dorsi-ventral curvature with the sides relatively straight and parallel for most of spore length. *2, 6, 8, 9, 10, 11, 12*
   c. Macroconidia very thin, needle-like, with thin walls. *1, 5, 10*
   d. Macroconidia relatively stout with a marked dorsi-ventral curvature. *8*
   e. Some macroconidia spindle-shaped. Spindle-shaped macroconidia are formed only in the aerial mycelium and may be produced on monophialides or polyphialides. *5, 6, 9*

3. Shape of basal and apical cells
   a. Basal cell not distinct or papillate. Not distinctly foot-shaped. *1, 3, 4*
   b. Basal cell distinctly foot-shaped or notched. *2, 4, 5, 6, 7, 8, 9, 10, 11, 12*
   c. Apical cell extended and whip-like. *7*

C. Microconidia from aerial mycelium

1. **Present or absent**
   a. Present and abundant. 2, 4, *7* (*F. scirpi*), *8* (*F. bactridioides*), *9,* 10, 11, 12
   b. Absent or sparse (i.e. less than 1/10 of conidia present in low power compound microscope field). 1, 3, 5, 6, *7, 8, 9*

2. **In chains or false heads**
   a. In chains and false heads. 2, 10
   b. In false heads only. 4, 11, 12

3. **Shape**
   a. Oval to egg-shaped (ovoid) to kidney-shaped (reniform) to fusiform. 2, *4, 7* (*F. scirpi*), 8 (*F. bactridioides*), *10,* 11, 12
   b. Globose (napiform). *4* (*F. poae*), *10* (*F. anthophilum*)

D. Conidiophores

1. **Type**
   a. Monophialides only (may produce either macroconidia or microconidia). 1, 2, 3, *4,* 5, *6, 7,* 8, 9, *10,* 11, 12
   b. Monophialides (may produce either macroconidia or microconidia) and polyphialides (produce only spindle-shaped macroconidia and microconidia). *4, 6, 7* (*F. scirpi*), *10*

E. Chlamydospores

1. **Present or absent**
   a. Present. *1,* 4, 6, 7, 8, 9, 11, 12
   b. Absent. *1,* 2, 3, 5, 10

2. **Arrangement**
   a. Single or in pairs. 1, *4,* 9, 11, 12
   b. Long chains or large clumps of more than three cells. *4,* 6, 7, 8

Synoptic Keys
to *Fusarium* Species

The synoptic keys to the species are more precise and for that reason may appear to contradict specific points in the synoptic key to the sections. This is unavoidable in this large and very diverse group. Where contradictions occur, the synoptic key to the species should be followed. The most complete descriptions are to be found under the species descriptions. These descriptions and their accompanying photographs are the final authority for identification. As in the previous key, pionnotal and mycelial mutants are not included. Where such mutants are to be identified, it is preferable to go directly to the figures for direct comparison of conidium and conidiophore morphology. As an aid to the reader, the numbers used to identify each species in the synoptic key appear in front of the species name in the descriptions of species in Part III. Numbers printed in italics indicate species included in more than one description under a given characteristic.

Species in the Sections Eupionnotes, Spicarioides, and Arachnites

Section Eupionnotes (1)
1. *F. aquaeductuum*
2. *F. merismoides*
3. *F. dimerum*

Section Spicarioides (2)
   4. *F. decemcellulare*
Section Arachnites (3)
   5. *F. nivale*

Cultural characteristics are based on 10- to 14-day-old cultures grown on PDA slants. The descriptions of conidia, conidiophores, and chlamydospores are based on 10- to 14-day-old cultures grown on CLA plates. All cultures were grown from single conidia from sporodochial isolates and, when possible, from isolates recently obtained from substrates in nature. Mycelial and pionnotal mutants are not included in the descriptions. The illustrations of species should be used for direct comparison of conidium and conidiophore morphology when attempting to identify mutants. Chlamydospores may form slowly in culture, often requiring one month or more, and both the CLA and the PDA preparations should be checked for the presence of chlamydospores.

SPECIES IN SECTIONS 1,2,3

A. Cultural Characteristics

1. Rate of growth
a. Growth slow, less than 2 to 3 cm diameter after 10 days. **1, 2, 3**
b. Growth relatively slow, less than 7 cm diameter after 10 days. **4, 5**

2. Aerial mycellium present or absent
a. Present. **4, 5**
b. Absent and colony surface slimy and yeast-like. **1, 2, 3**

3. Color of aerial mycelium
a. White. **4**
b. White to light orange. **5**

4. Color of colony from below. Color may diffuse into agar
a. Colorless to cream. **1, 2, 3**
b. Tan to carmine red. **4**
c. Cream to pale or bright orange. **5**

5. Color of spore masses (sporodochia)
a. Spore masses distinct, cream to yellow. **4**
b. Spore masses distinct, pale to bright orange. **5**
c. Spore masses making up entire colony surface, cream, tan, or orange. **1, 2, 3**

B. Macroconidia from sporodochia

1. Size
a. Small, 1–2 septate. *3,* **5**
b. Small to moderately large, 3–7 septate. **1, 2,** *3*
c. Very large, 9 septate or more. **4**

2. Shape
a. Spores straight with sides parallel for most of their length or slightly curved. *1, 3, 5*
b. Spores curved. *1, 3, 5*
c. Spores cylindrical to robust with the sides parallel for most of their length. **2, 4**
d. Spores thin, needle-like. *1*

3. Shape of basal and apical cells
a. Pointed, basal cell not distinctly notched. *1, 2, 3,* **5**
b. Blunt, or with distinctly shaped basal and apical cells. Basal cell distinctly notched. *1, 3,* **4**

C. Microconidia from aerial mycelium

1. Present or absent
a. Present. **4**
b. Absent. **1, 2, 3, 5**

2. In chains or false heads
a. In chains and false heads. **4**

3. Shape
a. Oval and 0–1 septate. **4**

D. Conidiophores

1. Type
a. Monophialides only (may produce either macroconidia or microconidia). **1, 2, 3, 4, 5**

E. Chlamydospores

1. Present or absent
a. Present. **2, 3**
b. Absent. **1, 4, 5**

SPECIES IN SECTIONS 1,2,3

Species in the Sections Sporotrichiella, Roseum, Arthrosporiella, Gibbosum, and Discolor

Section Sporotrichiella (4)
6. *F. poae*
7. *F. tricinctum*
8. *F. sporotrichioides*
9. *F. chlamydosporum*

Section Roseum (5)
10. *F. graminum*
11. *F. avenaceum* (including *F. arthrosporioides*)

Section Arthrosporiella (6)
12. *F. semitectum*
13. *F. camptoceras*

Section Gibbosum (7)
14. *F. equiseti* (including *F. scirpi* var. *compactum, F. scirpi* var. *filiferum)*
15. *F. scirpi*
16. *F. acuminatum*
17. *F. longipes*

Section Discolor (8)
18. *F. heterosporum*
19. *F. reticulatum*
20. *F. sambucinum* (including *F. bactridioides*)
21. *F. culmorum*
22. *F. graminearum*
23. *F. crookwellense*

Cultural characteristics are based on 10- to 14-day-old cultures grown on PDA slants. The descriptions of conidia, conidiophores, and chlamydospores are based on 10- to 14-day-old cultures grown on CLA plates. All cultures were grown from single conidia from sporodochial isolates and, when possible, from isolates recently obtained from substrates in nature. Mycelial and pionnotal mutants are not included in the descriptions. The illustrations of species should be used for direct comparison of conidium and conidiophore morphology when attempting to identify mutants. The presence or absence of microconidia and their shape should be checked on both CLA and PDA, especially in species in the section Sporotrichiella. Chlamydospores may form slowly in culture, often requiring one month or more, and both the CLA and the PDA preparations should be checked for the presence of chlamydospores.

A. Cultural characteristics

1. Rate of growth
a. Relatively slow, less than 7 cm diameter after 10 to 14 days. **19**
b. Growth rapid, more than 7 cm diameter after 10 to 14 days. **6, 7, 8, 9, 10, 11, 12, 13, 14, 15, 16, 17, 18, 20, 21, 22, 23**

2. Color of aerial mycelium
a. White. **6, 7, *8, 9,* 10, *11, 14, 15,* 16, 17, 18, 19, *20,* 21, *22, 23***
b. Tan, brown. ***8, 9, 11,* 12, 13, *14, 15, 20, 22, 23***

3. Color of colony from below. Colors may diffuse into agar
a. Tan to brown. ***8, 9,* 10, *11,* 12, 13, *14,* 15, *16, 17,* 18, *19, 20***
b. Carmine red. **6, 7, *8, 9, 11, 14*** (*F. scirpi* var. *compactum*), ***16, 17, 19, 20,* 21, 22, 23**

4. Color of spore masses (sporodochia)
a. Spores in light orange to orange masses scattered over the surface of the colony. **7, 8, 12, 14, 15, *20***
b. Spores in reddish-brown masses concentrated in the center of the colony. ***11, 16, 21, 22, 23***
c. Spores in orange masses concentrated in the center of the colony. **10, *11, 16,* 17, 18, 19, *20, 21, 22, 23***

B. Macroconidia from sporodochia

1. Shape
a. Thin, needle-like, with thin walls. **10, *11***
b. Stout, with thick walls. ***20, 21, 22, 23***

c. Curved but with the walls mainly parallel through most of their length. **6, 7, 8, 9, 18, *20, 21, 22***
d. Strong dorsi-ventral curvature (hunch-backed). **14, 15, 16, 17, 19, *23***
e. Spindle-shaped and produced on monophialides or polyphialides in the aerial mycelium. ***11,* 12, 13**

2. Shape of basal and apical cells
a. Basal cell distinctly notched or foot-shaped (pedicellate). **6, 7, 9, 10, 11, *14, 15, 16, 17,* 18, 19, *20, 21, 22, 23***
b. Basal cell not distinctly foot-shaped. **8, *21***
c. Basal cell with a papilla. **12, 13**
d. Apical cell, cone-like, elongated or whip-like. ***14, 15, 16, 17***
e. Apical cell nipple-like, sometimes strongly curved as a beak. ***20, 21, 22, 23***

C. Microconidia from aerial mycelium

1. Present or absent
a. Scarce to none. **10, 11, 12, 13, *14,* 16, 17, 18, 19, *20,* 21, 22, 23**
b. Fairly abundant. ***14,* 15, *20*** (*F. bactridioides*)
c. Abundant. **6, 7, 8, 9**

2. Shape
a. Oval, ellipsoidal, comma-shaped, and club-shaped. **14, 15, 20**
b. Globose (napiform). **6**
c. Lemon-shaped, pear-shaped, and spindle-shaped. **7, 8**
d. Mainly spindle-shaped. **9**

D. Conidiophores

1. Type
   a. Monophialides only (may produce either macroconidia or microconidia). **6, 7, 10, 11, 14, 16, 17, 18, 19, 20, 21, 22, 23**
   b. Monophialides (may produce either macroconidia or microconidia) and polyphialides (produce only microconidia and spindle-shaped macroconidia in the aerial mycelium). **8, 9, 12, 13, 15**

E. Chlamydospores

1. Present or absent
   a. Present. **6, 7, 8, 9, 12, 13, 14, 15, 16, 17, 18, 19, 20, 21, 22, 23**
   b. Absent. **10, 11**

S
P
E
C
I
E
S

IN

SECTIONS

4,5,6,
7,8

Species in the Sections Lateritium, Liseola, Elegans, Martiella, and Ventricosum

Section Lateritium (9)
24. *F. lateritium* (including *F. stilboides*)
Section Liseola (10)
25. *F. moniliforme*
26. *F. proliferatum*
27. *F. subglutinans*
28. *F. anthophilum*
Section Elegans (11)
29. *F. oxysporum* (including *F. udum* and *F. redolens*)
Sections Martiella and Ventricosum (12)
30. *F. solani*

Cultural characteristics are based on 10- to 14-day-old cultures grown on PDA slants. The descriptions of conidia, conidiophores, and chlamydospores are based on 10- to 14-day-old cultures grown on CLA plates. All cultures were grown from single conidia from sporodochial isolates and, when possible, from isolates recently obtained from substrates in nature. Mycelial and pionnotal mutants are not included in the descriptions. The illustrations of species should be used for direct comparison of conidium and conidiophore morphology when attempting to identify mutants. Chlamydospores may form slowly in culture, often requiring one month or more, and both the CLA and the PDA preparations should be checked for the presence of chlamydospores.

SPECIES IN SECTIONS 9,10, 11,12

A. **Cultural characteristics.**

1. **Rate of growth**
   a. Relatively slow, less than 7 cm diameter after 10 to 14 days. **24**
   b. Growth rapid, greater than 7 cm diameter after 10 to 14 days. **25, 26, 27, 28, 29, 30**

2. **Color of aerial mycelium**
   a. White. *24, 25, 26, 27, 28, 29, 30*
   b. Tan to carmine red. *24*
   c. Light purple. *25, 26, 27, 28, 29*

3. **Color of colony from below. Color may diffuse into agar**
   a. Carmine red. *24*
   b. Tan to orange. *24, 25, 26, 27, 28, 29, 30*
   c. Light to dark purple pigment diffusing into agar. *25, 26, 27, 28, 29, 30*

4. **Color of spore masses (sporodochia)**
   a. Tan to orange. **24, 25, 26, 27, 28, *29***
   b. Cream. *29, 30*
   c. Blue-green to blue. *30*

B. **Macroconidia from sporodochia**

1. **Size**
   a. Long. *25, 26, 27, 28, 30*
   b. Short. *25, 26, 27, 28, 29, 30*

2. **Shape**
   a. Thin, with thin walls. *25, 26, 27, 28*
   b. Relatively thin, with thin walls. *25, 26, 27, 28, 29*
   c. Stout with thick walls, and walls parallel through most of their length. **24, 30**

3. **Shape of basal and apical cells**
   a. Apical cell with distinct beak or sharply curved. **24**
   b. Apical cell not distinctly shaped. **30**

C. **Microconidia from aerial mycelium**

1. **Present or absent**
   a. Sparse. *24*
   b. Abundant. *24,* 25, 26, 27, 28, 29, 30

2. **In chains or false heads**
   a. In chains and false heads. **25, 26**
   b. In false heads only. **24, 27, 28, 29, 30**

3. **Shape**
   a. Oval to kidney-shaped (reniform) and clavate. **24, 25, 26, 27, 29, 30**
   b. Oval, pear-shaped or globose (napiform). **28**

D. **Conidiophores**

1. **Type**
   a. Medium length monophialides producing microconidia. **24**
   b. Short monophialides producing microconidia. **29**
   c. Long monophialides producing microconidia. **25, 30**
   d. Monophialides and polyphialides producing microconidia. **26, 27, 28**

E. **Chlamydospores**

1. **Present or absent**
   a. Absent. **25, 26, 27, 28**
   b. Sparse. **24**
   c. Abundant. **29, 30**

PART III
Descriptions and Illustrations of *Fusarium* Species

Descriptions and Illustrations of Well-Documented *Fusarium* Species

Descriptions and photomicrographs of conidia, coniophores, and chlamydospores are of material taken from cultures grown on carnation-leaf agar, prepared as described on pp. 13–14, for 5 to 14 days under the conditions described on pp. 15–16.

Descriptions of colony morphology, including color, are based on cultures grown for 10 to 14 days on potato dextrose agar slants, prepared as described on pp. 12–13, and under the conditions described on pp. 15–16.

The designations given in brackets after each species epithet or synonym refer to the taxonomic system within which that species name is used. The following abbreviations are used in those designations:

W&R = Wollenweber and Reinking
S&H = Snyder and Hansen
M&C = Messiaen and Cassini

G = Gerlach
B = Booth
J = Joffe

A statement at the end of a species description noting that the species "has been reported to be toxigenic" indicates that the species has been described in the literature as toxigenic and that a complete discussion of its toxigenicity is included in the companion volume, *Toxigenic* Fusarium *Species: Identity and Mycotoxicology,* by W.F.O. Marasas, P.E. Nelson, and T.A. Toussoun.

1. SECTION EUPIONNOTES
1. *Fusarium aquaeductuum* (Radlk. & Rabenh.) Lagerh. [W&R,G,B,J]
   Syn.: *F. episphaeria* (Tode) Snyd. & Hans. pro parte [S&H,M&C]

Conidia (Figs. 1, 2a). Microconidia are absent. Macroconidia are abundant, delicate, thin-walled and generally crescent-shaped with pointed ends. The basal cell may or may not be foot-shaped.
Conidiophores (Fig. 2b). Unbranched and branched monophialides.
Chlamydospores. Absent.
Perfect state. *Nectria episphaeria* (Tode) Fr. (5).
Colony morphology. On PDA, *F. aquaeductuum* is very slow growing, forming yeast-like colonies that appear to be wet and have little visible aerial mycelium. Colony color varies from white to orange, with the latter color being the most common. The undersurface is generally the color of the upper surface.

Most distinguishing characteristics. The colony morphology on PDA, and the shape of the macroconidia.
Affinities. The species of Wollenweber and Reinking's section Eupionnotes (*F. aquaeductuum, F. merismoides, F. dimerum*) are similar to each other in culture, and are quite different in colony morphology on PDA from all the other *Fusarium* species.

Wollenweber and Reinking (76) separated one variety, *F. aqueductuum* var. *medium,* because it had longer macroconidia. We have included this variety in *F. aquaeductuum.*

Fusarium aquaeductuum is stable in culture.

Fusarium aquaeductuum is cosmopolitan.

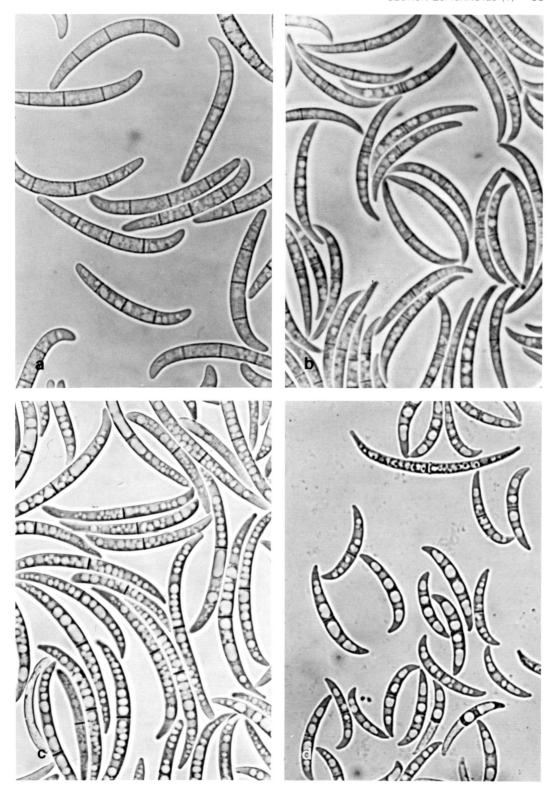

Figure 1. *Fusarium aquaeductuum:* a–d, macroconidia (X1000).

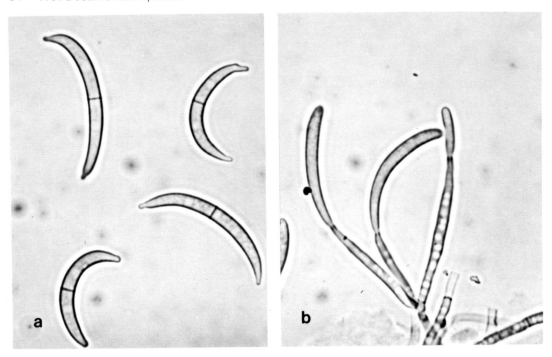

Figure 2. *Fusarium aquaeductuum:* a, macroconidia; b, conidiophores (monophialides) (X1000).

1. SECTION EUPIONNOTES
2. *Fusarium merismoides* Corda [W&R,G,B,J]
Syn.: *F. episphaeria* (Tode) Snyd. & Hans. pro parte [S&H,M&C]

Conidia (Fig. 3). Microconidia are absent. Macroconidia are abundant, slightly curved and generally more stout than in *F. aquaeductuum,* with a blunt apical cell that is sometimes hooked. The basal cell is not foot-shaped or notched.
Conidiophores (Fig. 4a,b). Unbranched and branched monophialides.
Chlamydospores (Fig. 4c–e). Present, and sometimes occur in chains.
Perfect state. None known.
Colony morphology. On PDA, *F. merismoides* is very slow growing, forming yeast-like colonies that appear to be wet and have little visible aerial mycelium. Colony color varies from white to orange with the latter color being the most common. The undersurface is generally the color of the upper surface.

Most distinguishing characteristics. The colony morphology on PDA and the shape of the macroconidia.
Affinities. The species of Wollenweber and Reinking's (76) section Eupionnotes (*F. aquaeductuum, F. merismoides, F. dimerum*) are similar in culture and quite different in colony morphology on PDA from all other *Fusarium* species.

Wollenweber and Reinking (76) established two varieties, *F. merismoides* var. *chlamydosporale* and var. *crassum,* on the basis of the occurrence of larger macroconidia, more abundant chlamydospores, and larger chlamydospores. We have included these varieties in *F. merismoides.*

Fusarium merismoides is stable in culture.

Fusarium merismoides is cosmopolitan.

Fusarium merismoides has been reported to be toxigenic.

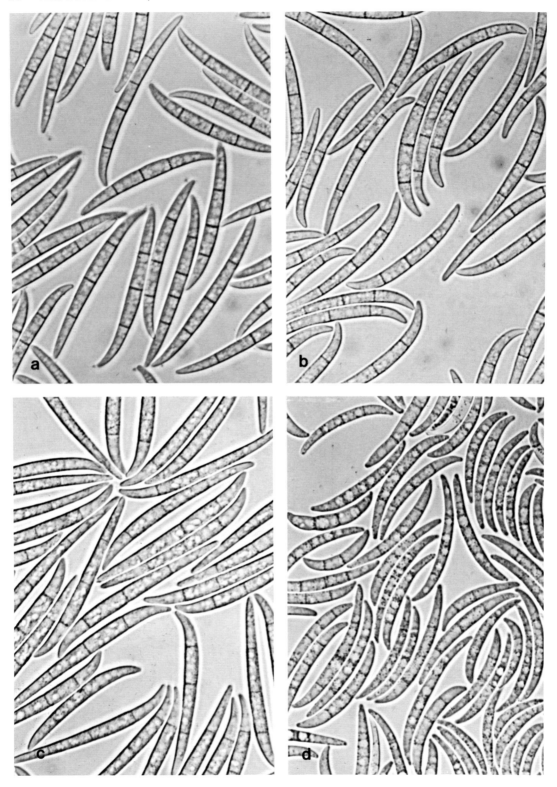

Figure 3. *Fusarium merismoides:* a–d, macroconidia (X1000).

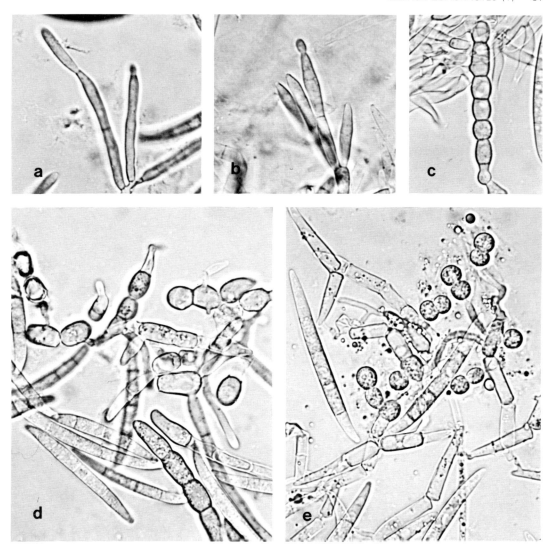

Figure 4. *Fusarium merismoides:* a, b, conidiophores (monophialides); c—e, chlamydospores (X1000).

1. SECTION EUPIONNOTES
3. *Fusarium dimerum* **Penzig** [W&R,G,B,J]
Syn.: *F. episphaeria* (Tode) Snyd. & Hans. pro parte [S&H,M&C]

Conidia (Fig. 5a–c). Microconidia are absent. Macroconidia are abundant, small, 1-2 septate, and resemble macroconidia of *F. nivale*. The apical cell may be hooked and the basal cell is blunt or slightly notched.
Conidiophores (Fig. 5d,e). Unbranched and branched monophialides.
Chlamydospores (Fig. 5f,g). Present but may be rare.
Perfect state. None known.
Colony morphology. On PDA, *F. dimerum* is very slow growing, forming yeast-like colonies that appear to be wet and have little visible aerial mycelium. Colony color varies from white to orange with the latter color generally being most common. The undersurface is generally the color of the upper surface.

Most distinguishing characteristics. Colony morphology on PDA, and the shape of the macroconidia. While the shape of the macroconidia of *F. dimerum* may resemble those of *F. nivale*, colony appearance on PDA is quite different and *F. dimerum* grows more slowly than *F. nivale*.
Affinities. The species of Wollenweber and Reinking's (76) section Eupionnotes (*F. aquaeductuum, F. merismoides, F. dimerum*) are similar in culture and quite different in colony morphology on PDA, from all other *Fusarium* species.

Wollenweber and Reinking (76) recognize three varieties, *F. dimerum* var. *nectrioides,* var. *pusillum,* and var. *violaceum,* based on size of macroconidia and colony color; we include all three in *F. dimerum.*

Fusarium dimerum is stable in culture.

Fusarium dimerum is cosmopolitan.

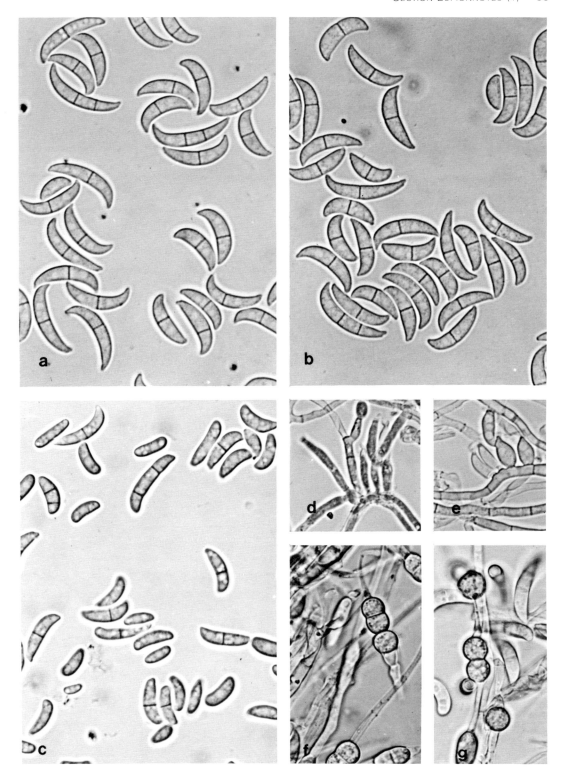

Figure 5. *Fusarium dimerum:* a–c, macroconidia; d, e, conidiophores (monophialides); f, g, chlamydospores (X1000).

2. SECTION SPICARIOIDES

4. *Fusarium decemcellulare* Brick [W&R,G,B,J]

Syn.: *F. rigidiuscula* (Brick) Snyd. & Hans. [S&H,M&C]

Conidia (Fig. 6a,b,f,g). Microconidia are abundant, oval, 0-1 septate, relatively large, and are formed in false heads and in very long chains (Fig. 6e). Macroconidia are very large with thick walls and are generally 9 septate. The dorsal and ventral surfaces of the macroconidia are equally but only slightly curved over most of the length of the macroconidia. The apical cell and the basal cell are similar in shape, but the basal cell is sometimes more prominently notched.

Conidiophores (Fig. 6c,d). Unbranched and branched monophialides.

Chlamydospores. Absent.

Perfect state. *Calonectria rigidiuscula* (Berk. & Br.) Sacc. (5).

Colony morphology. On PDA, the aerial mycelium is white, and prominent cream to yellow sporodochia form rapidly. Droplets of exudate, which form on the sporodochia, are frequently so liquid that moisture flows over the culture. The undersurface is usually carmine red.

Most distinguishing characteristics. The striking coloration of the colony on PDA. This species has the largest macroconidia of *Fusarium* species commonly seen and large microconidia in long chains.

Affinities. *Fusarium decemcellulare* is the only species in Wollenweber and Reinking's section Spicarioides. It is unlike any other *Fusarium* species.

Fusarium decemcellulare is fairly stable in culture, but may mutate to pionnotal or mycelial forms. In both cases the prominent sporodochia are absent or much reduced. In pionnotal forms the microconidia in chains are rare but the typical large macroconidia are plentiful. In mycelial forms the macroconidia are not as abundant but the microconidia formed in chains are generally still prominent.

Fusarium decemcellulare is usually found in tropical and sub-tropical areas.

Fusarium decemcellulare has been reported to be toxigenic.

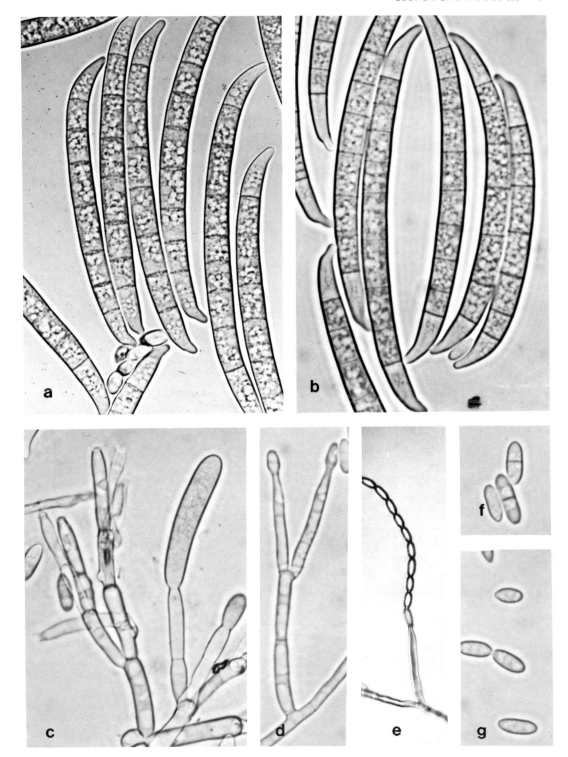

Figure 6. *Fusarium decemcellulare:* a, b, macroconidia; c, macroconidiophores (monophialides); d, microconidiophores (monophialides); e, chain of microconidia on a monophialide; f, g, microconidia (a–d, f, g X1000; e X400).

3. SECTION ARACHNITES
5. *Fusarium nivale* (Fr.) Ces. [W&R,G,B,J]
Syn.: *F. nivale* Fr. emend. Snyd. & Hans. pro parte [S&H,M&C]

Conidia (Fig. 7a–c). Microconidia are absent. Macroconidia are small, 1–3 septate, curved, and with blunt terminal cells that are indistinguishable from one another. The basal cell may be notched but is never foot-shaped.
Conidiophores (Fig. 7d). Unbranched and branched monophialides.
Chlamydospores. Absent.
Perfect state. *Monographella nivalis* (Schaffnit) Müller (5).
Colony morphology. On PDA, *F. nivale* is relatively slow growing and has felted aerial mycelium. The colonies are white at first and become pink to orange as the culture ages. Discrete orange sporodochia may appear as the culture ages. The undersurface is colorless to light orange.

Most distinguishing characteristics. The shape of macroconidia and the fact that optimum growth occurs at cool temperatures of 18 C or less.
Affinities. None. *Fusarium nivale* is the only species that grows and sporulates best at temperatures of 18 C or lower.

Wollenweber and Reinking (76) distinguished the variety *F. nivale* var. *majus* on the basis of larger macroconidia; we include it in *F. nivale*. Recently Gams and Müller (15) transferred *F. nivale* from the genus *Fusarium* to the new genus *Gerlachia* as *G. nivalis* (Ces. ex Sacc.) W. Gams & E. Müller. These authors demonstrated the presence of annellate conidiogenous cells in this species, and this character, together with the fact that its teleomorph, *Monographella nivalis,* belongs in the Amphisphaeriaceae, distinguishes *F. nivale* from all other *Fusarium* species. However, we are retaining *F. nivale* in *Fusarium* until more isolates have been examined and until more information becomes available on the validity of annellate conidiogenous cells as a generic character. The cultures we have examined under the scanning-electron microscope have not shown clearly the annellations on the phialides reported by Gams and Müller (15).

Fusarium nivale is stable in culture.

Fusarium nivale is cosmopolitan but not found in sub-tropical or tropical regions.

Fusarium nivale has been reported to be toxigenic.

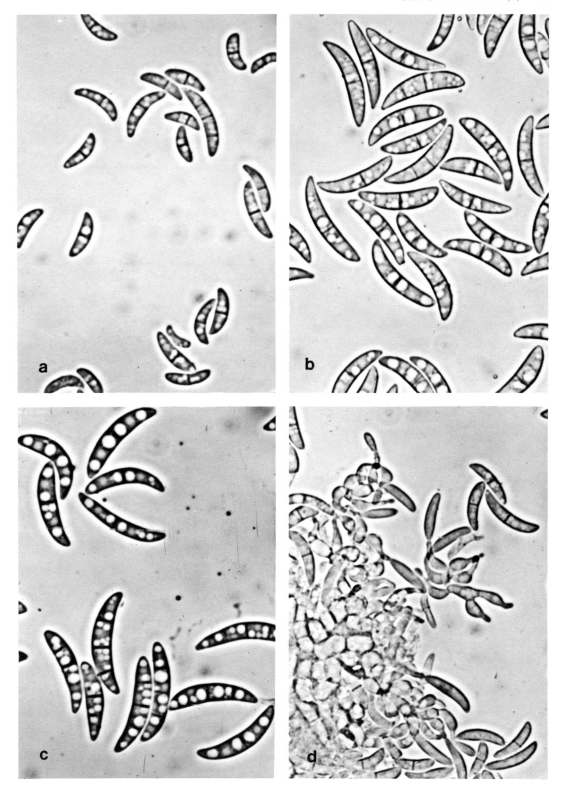

Figure 7. *Fusarium nivale:* a–c, macroconidia; d, conidiophores (monophialides) (X1000).

4. SECTION SPOROTRICHIELLA
6. *Fusarium poae* (Peck) Wollenw. [W&R,G,B,J]
Syn. *F. tricinctum* Corda emend. Snyd. & Hans. pro parte [S&H,M&C]

Conidia (Figs. 8, 9). Microconidia are abundant, globose or oval in shape, 0–1 septate, and often have a distinct papilla. Macroconidia are generally rare, typically sickle-shaped, 3–5 septate, and have a foot-shaped basal cell.
Conidophores (Fig. 9). Unbranched and branched monophialides. Microconidiophores are short and fat, almost globose, and quite distinctive.
Chlamydospores. Occur infrequently and may be in clumps or chains.
Perfect state. None known.
Colony morphology. On PDA, growth is rapid, with dense aerial mycelium that is white to pink in color. Sporodochia are rare. The undersurface is usually carmine red.

Most distinguishing characteristics. The abundant production of globose to oval microconidia borne on monophialides.
Affinities. The rapid, profuse mycelial growth and carmine red undersurface on PDA make *F. poae* quite similar in appearance to some species in the sections Arthrosporiella, Roseum, and Discolor. However, the abundant production of globose microconidia place it in the section Sporotrichiella of Wollenweber and Reinking (76).

Fusarium poae is stable in culture.

Fusarium poae is cosmopolitan.

Fusarium poae has been reported to be toxigenic.

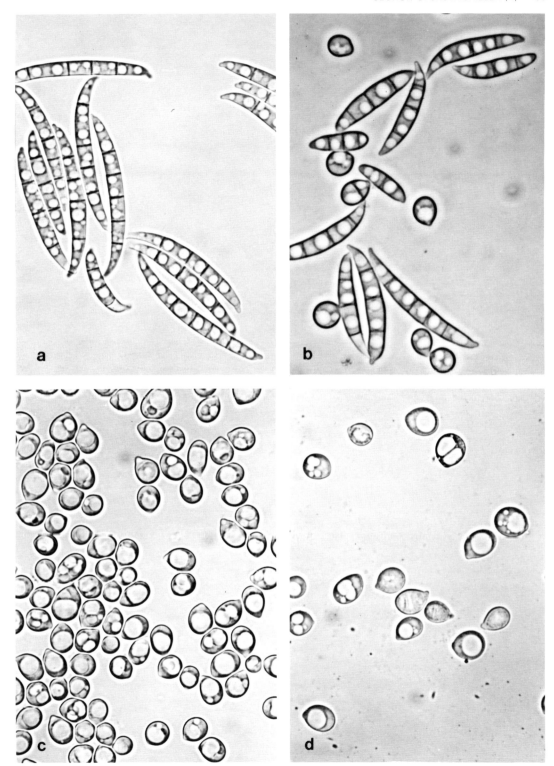

Figure 8. *Fusarium poae:* a, macroconidia; b, macroconidia and microconidia; c, d, microconidia (X1000).

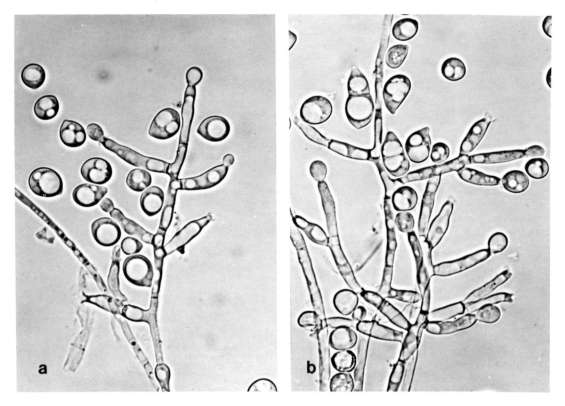

Figure 9. *Fusarium poae:* a, b, microconidia and microconidiophores (monophialides) (X1000).

4. SECTION SPOROTRICHIELLA
7. *Fusarium tricinctum* (Corda) Sacc. [W&R,G,B]

      Syn.: *F. sporotrichioides* Sherb. var. *tricinctum* (Corda) Raillo [J]
          F. tricinctum Corda emend. Snyd. & Hans. pro parte [S&H,M&C]

Conidia (Fig. 10). Microconidia are abundant and are lemon- to pear-shaped or spindle-shaped, 0–1 septate, and often have a papilla at the base. Macroconidia are also abundant, sickle-shaped, and are similar in shape to macroconidia of *F. reticulatum*. The basal cell is distinctly foot-shaped or notched.
Conidiophores (Fig. 11). Unbranched and branched monophialides.
Chlamydospores. Present and formed singly or in chains.
Perfect state. *Gibberella tricincta* El-Gholl, McRitchie, Schoulties, and Ridings (12) is possibly the perfect state of *F. tricinctum* sensu stricto.
Colony morphology. On PDA growth is rapid, and the aerial mycelium is dense and white, with orange sporodochia appearing as the culture ages. The undersurface is carmine red.

Most distinguishing characteristics. Abundant microconidia that are either lemon- to pear shaped or spindle-shaped and are borne on unbranched or branched monophialides. The absence of polyphialides separates *F. tricinctum* from *F. chlamydosporum* and *F. sporotrichioides*, while the presence of spindle-shaped microconidia separates it from *F. poae*.
Affinities. On PDA, *F. tricinctum* has the profuse mycelial growth and carmine red under-surface typical of some species in the sections Arthrosporiella, Roseum, and Discolor, but the presence of abundant lemon- or spindle-shaped microconidia indicates it is a member of Wollenweber and Reinking's (76) section Sporotrichiella.

It is important to examine the cultures on PDA and on CLA to find the two kinds of microconidia.

Fusarium tricinctum is stable in culture.

Fusarium tricinctum is cosmopolitan, but appears to be more common in temperate areas.

Fusarium tricinctum has been reported to be toxigenic.

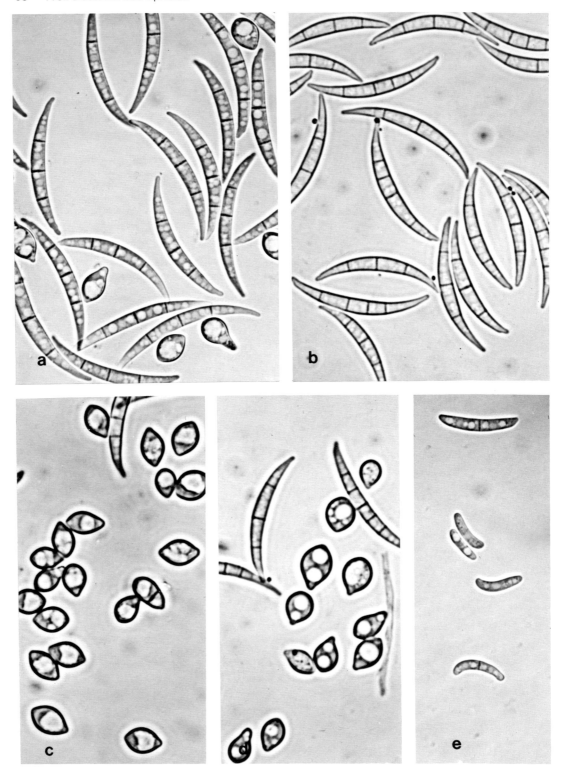

Figure 10. *Fusarium tricinctum:* a, macroconidia and microconidia; b, macroconidia; c–e, microconidia (X1000).

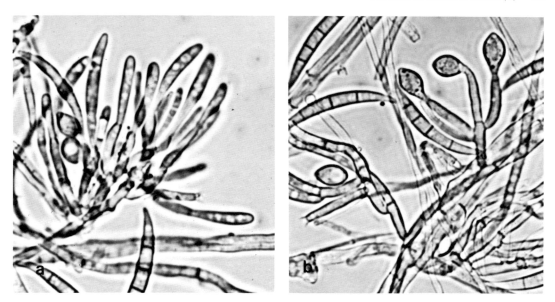

Figure 11. *Fusarium tricinctum:* a, macroconidiophores (monophialides); b, microconidiophores (monophialides) (X1000).

4. SECTION SPOROTRICHIELLA

8. *Fusarium sporotrichioides* Sherb. [W&R,G,B,J]

Syn.: *F. tricinctum* Corda emend. Snyd. & Hans. pro parte [S&H,M&C]

Conidia (Figs. 12, 13). Microconidia are abundant, and are oval to pear-shaped or spindle-shaped, 0–1 septate, often with a papilla at the base. Macroconidia are also abundant, sickle-shaped, 3–5 septate, and occasionally similar in form to macroconidia of *F. reticulatum.* The basal cell is not distinctly foot-shaped or notched.

Conidiophores (Fig. 14a–c). Unbranched and branched monophialides and polyphialides bearing microconidia, and unbranched and branched monophialides bearing macroconidia.

Chlamydospores (Fig. 14d,e). Present and abundant and formed singly, in chains, or in clumps.

Perfect state. None known.

Colony morphology. On PDA, growth is rapid and the aerial mycelium is dense and white, with orange sporodochia appearing as the culture ages. The undersurface is carmine red.

Most distinguishing characteristics. Polyphialides bearing two kinds of microconidia either oval to pear-shaped or spindle-shaped. The presence of oval to pear-shaped microconidia separates *F. sporotrichioides* from *F. chlamydosporum,* while the presence of polyphialides separates it from *F. poae* and *F. tricinctum.*

Affinities. As with *F. poae* and many clones of *F. chlamydosporum,* the cultures of *F. sporotrichioides* have the profuse mycelial growth and carmine red undersurface typical of some species in the sections Arthrosporiella, Roseum, and Discolor. However, the abundance of microconidia identifies it as a member of Wollenweber and Reinking's (76) section Sporotrichiella.

It is important to inspect cultures grown on PDA as well as on CLA to find the two kinds of microconidia that are diagnostic for this species.

Wollenweber and Reinking (76) recognize a variety, *F. sporotrichioides* var. *minus,* based on a lack of carmine red color on the undersurface, fewer sporodochia, and wider macroconidia. We include this variety in *F. sporotrichioides.*

Fusarium sporotrichioides is stable in culture.

Fusarium sporotrichioides occurs in cold to cold-temperate areas.

Fusarium sporotrichioides has been reported to be toxigenic.

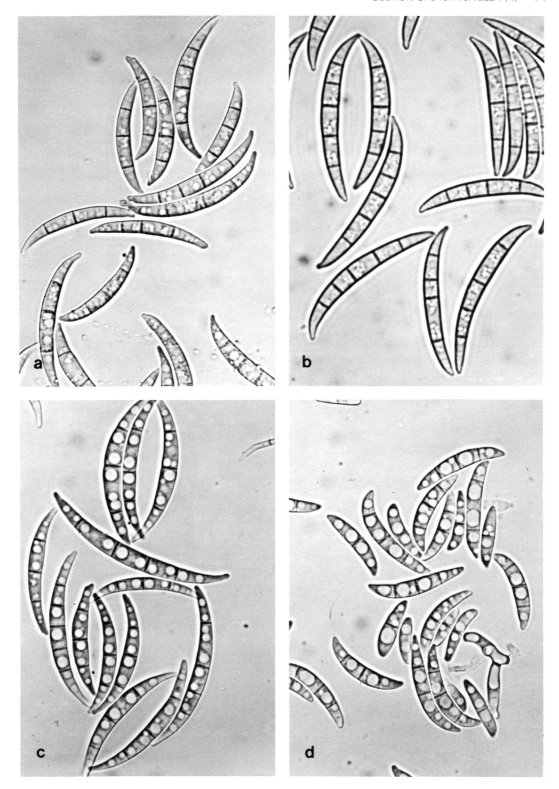

Figure 12. *Fusarium sporotrichioides:* a–d, macroconidia (X1000).

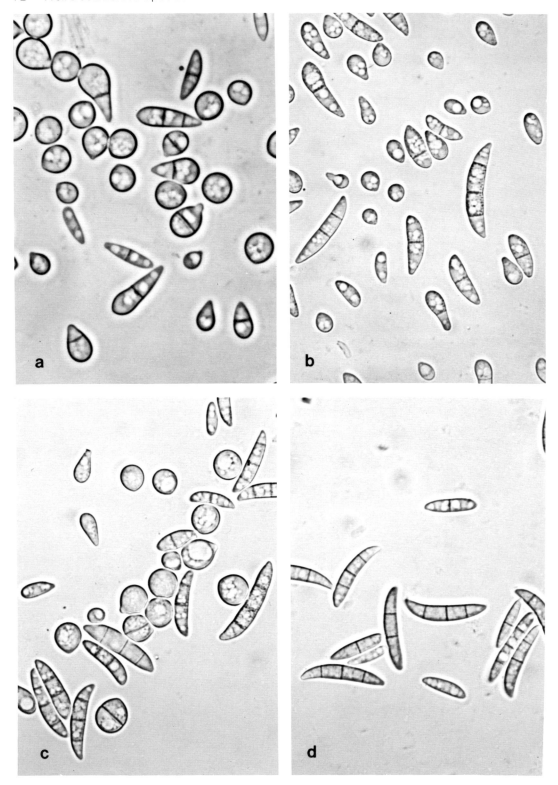

Figure 13. *Fusarium sporotrichioides:* a–d, microconidia (X1000).

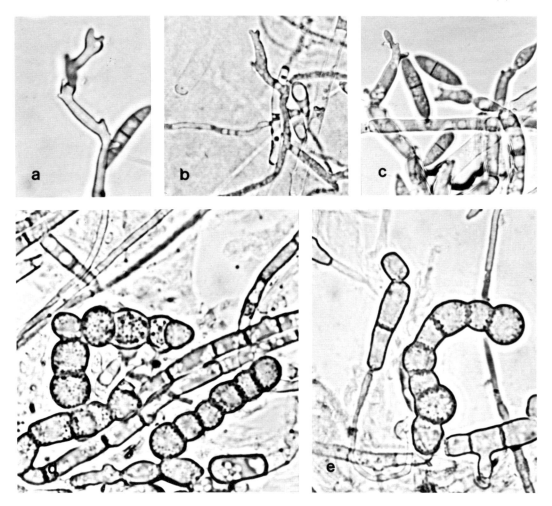

Figure 14. *Fusarium sporotrichioides:* a–c, polyphialides; d e, chlamydospores (X1000).

4. SECTION SPOROTRICHIELLA

9. *Fusarium chlamydosporum* Wollenw. & Reinking [W&R,G]

Syn.: *F. fusarioides* (Frag. & Cif.) Booth [B]

F. sporotrichioides Sherb. var. *chlamydosporum* (Wollenw. & Reinking) Joffe [J]

F. tricinctum Corda emend. Snyd. & Hans. pro parte [S&H,M&C]

Conidia (Fig. 15). Microconidia are abundant, 0–1 septate, mostly spindle-shaped but not globose. Macroconidia are generally rare, typically sickle-shaped, 3–5 septate, and have a basal cell that is foot-shaped.

Conidiophores (Fig. 16a–c). Microconidiophores are unbranched and branched monophialides or polyphialides. Macroconidiophores are unbranched and branched monophialides.

Chlamydospores (Fig. 16d–f). Present and abundant, rough walled, and occur in pairs, chains, or clumps. Chlamydospores may occur in such large quantities that they cause the mycelium to become brown in color.

Perfect state. None known.

Colony morphology. On PDA, growth is rapid with dense, white to pink to brown aerial mycelium. Sporodochia are rare. The undersurface is generally carmine red but may also be tan to brown.

Most distinguishing characteristics. The presence of spindle-shaped microconidia borne on polyphialides.

Affinities. As with *F. poae, F. chlamydosporum* has the profuse mycelial growth and frequently the carmine red undersurface of some species in the sections Arthrosporiella, Roseum, and Discolor. However, the abundance of microconidia identify it as a member of Wollenweber and Reinking's (76) section Sporotrichiella. On PDA, the non-carmine red clones, referred to as *F. chlamydosporum* var. *fuscum* Gerlach (19), may resemble *F. semitectum*.

Fusarium chlamydosporum is stable in culture.

Fusarium chlamydosporum is cosmopolitan, but appears to be more common in tropical and sub-tropical areas.

Fusarium chlamydosporum has been reported to be toxigenic.

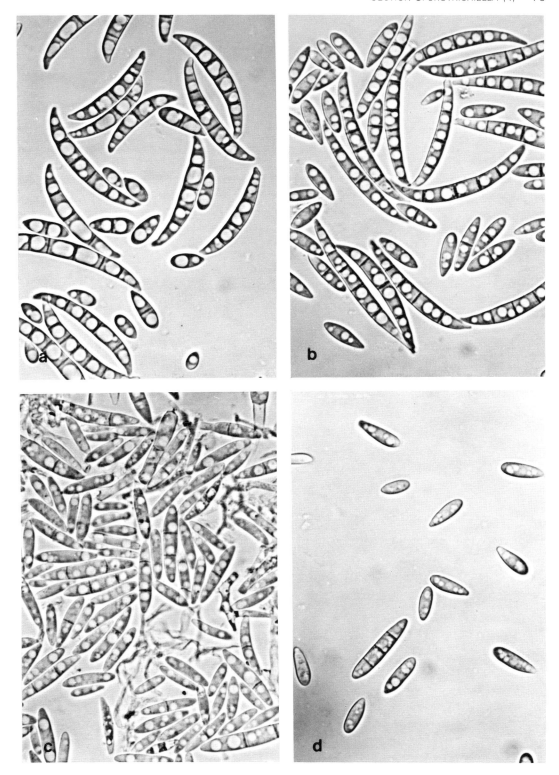

Figure 15. *Fusarium chlamydosporum:* a, b, macroconidia and microconidia; c, d, microconidia (X1000).

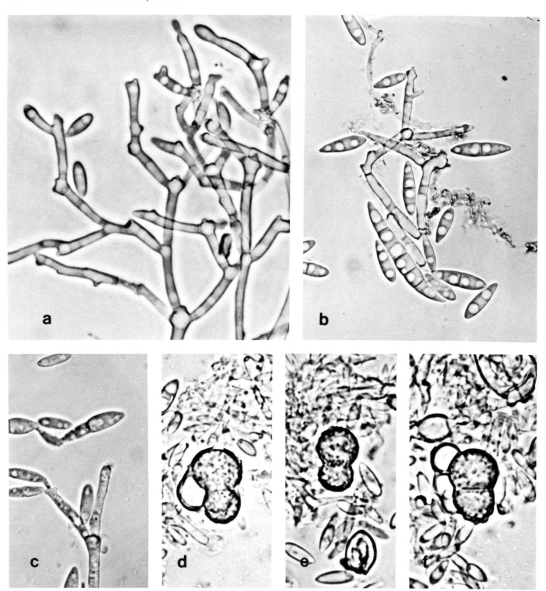

Figure 16. *Fusarium chlamydosporum:* a, polyphialides; b, c, polyphialides and microconidia; d–f, chlamydospores (X1000).

5. SECTION ROSEUM
10. *Fusarium graminum* Corda [W&R,G]

        Syn.: *F. heterosporum* Nees [B]

               F. roseum Lk. emend. Snyd. & Hans. pro parte [S&H,M&C]

               [Not mentioned by Joffe]

Conidia (Figs. 17, 18). Microconidia absent. Macroconidia are sickle-shaped, very slender, delicate, thin-walled, and have a foot-shaped basal cell. They are predominantly 3 septate, but some are 5 septate.

Conidiophores (Fig. 18). Unbranched and branched monophialides.

Chlamydospores. Absent.

Perfect state. None known.

Colony morphology. On PDA, growth is rapid and the aerial mycelium is dense and white to pink in color. Orange sporodochia develop as the culture ages. The undersurface is orange to tan.

Most distinguishing characteristics. The colony appearance on PDA and the morphology of the conidia produced on CLA. It resembles *F. heterosporum* in culture on PDA, with bright orange sporodochia that occur in fluffy white to pink mycelium, but the macroconidia on CLA are very thin and delicate and no chlamydospores are formed.

Affinities. *Fusarium graminum* resembles *F. heterosporum* in colony morphology on PDA. Conidia produced on CLA show a close resemblance to those of *F. avenaceum,* the other species in Wollenweber and Reinking's (76) section Roseum. Like *F. avenaceum* it does not form chlamydospores.

Mycelial mutants of *F. graminum* lose the conspicuous sporodochia, and the colonies have a featureless white appearance.

Fusarium graminum is cosmopolitan.

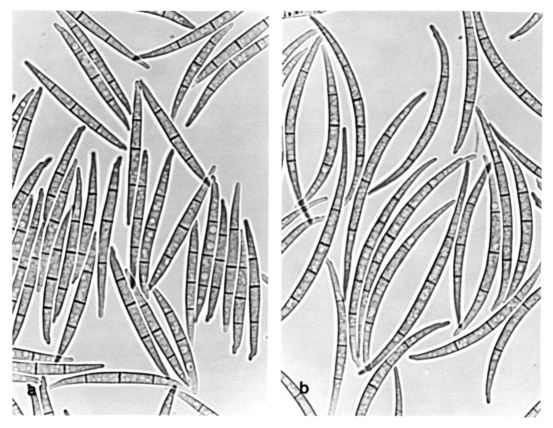

Figure 17. *Fusarium graminum:* a, b, macroconidia (X1000).

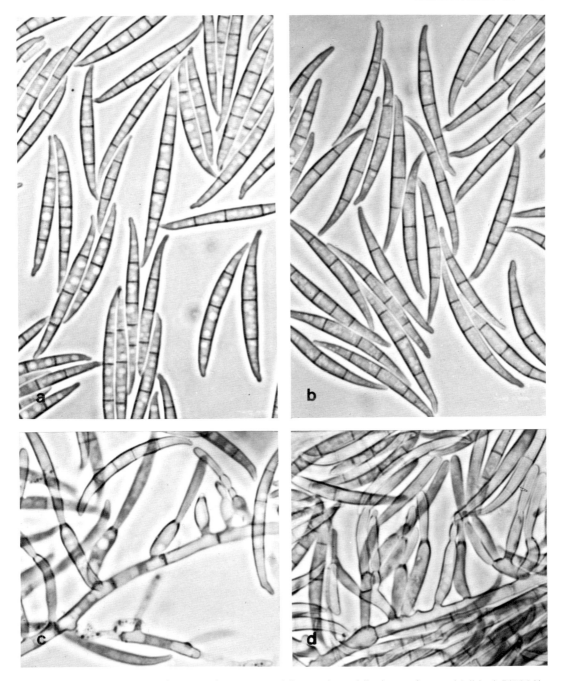

Figure 18. *Fusarium graminum:* a, b, macroconidia; c, d, conidiophores (monophialides) (X1000).

5. SECTION ROSEUM

11. *Fusarium avenaceum* (Fr.) Sacc. [W & R,G,B,J]

> Syn.: *F. roseum* Lk emend. Snyd. & Hans. 'Avenaceum' [S&H]
>
> *F. roseum* Lk. emend. Snyd. & Hans. var. *avenaceum* (Sacc.) Snyd. & Hans. [M&C]

Conidia (Figs. 19, 20a,b). Microconidia are generally scarce (see comments on *F. arthrosporioides* below). Macroconidia are very long, slender, and thin-walled, with an apical cell that is elongate and may be bent. The basal cell is foot-shaped or notched.
Conidiophores (Fig. 20c–e). Unbranched and branched monophialides.
Chlamydospores. Absent.
Perfect state. *Gibberella avenacea* Cook (11).
Colony morphology. On PDA, growth is rapid, the aerial mycelium is dense and white but frequently varies in color from tan to reddish-brown. Orange sporodochia may be present as the culture ages. The undersurface coloration varies from tan to carmine red to dark brown.

Most distinguishing characteristics. Macroconidia produced on CLA are long, slender, and generally more than 3 septate. No chlamydospores are produced.
Affinities. *Fusarium avenaceum* has a rapid profuse mycelial growth on PDA, and when the undersurface is carmine red it may resemble cultures of *F. graminearum, F. culmorum,* and *F. crookwellense.* The macroconidia produced on CLA are slender and thin-walled, similar to those produced by *F. graminum* but longer and generally more than 3 septate. This fungus is represented by many clones which differ considerably in colony appearance and morphology of the macroconidia, although the macroconidia are always slender and thin-walled. Some clones may resemble clones of *F. acuminatum* in growth and colonial morphology to such an extent that this distinction is one of the most difficult we encounter. A major difference is in chlamydospore production, since members of the section Roseum (*F. avenaceum, F. graminum*) do not form chlamydospores, while *F. acuminatum* (section Gibbosum) does. However, chlamydospores form very slowly in some clones of *F. acuminatum.*

Wollenweber and Reinking (76) used *F. arthrosporioides* Sherb. for those clones having oval-shaped, 0–3 septate, elongate spindle-shaped microconidia, along with macroconidia that are slender and thin-walled like those in *F. avenaceum.* We have seen some of these clones, but since the production of microconidia in *F. avenaceum* clones is not a clear-cut characteristic but rather occurs as a gradation between two extremes, we have elected to provisionally place *F. arthrosporioides* in *F. avenaceum,* until such time when it can be determined if a clear-cut demarcation between the two exists. It should be noted that *F. arthrosporioides* sensu Booth (4) is quite distinct from Wollenweber and Reinking's (76) concept of *F. arthrosporioides.*

Wollenweber and Reinking (76) also have one forma and two varieties, *F. avenaceum* var. *pallens* and var. *volutum,* within *F. avenaceum,* separated on the basis of the color of the stroma and the sporodochia. We have included those in *F. avenaceum. Fusarium detonianum* Sacc., in which Wollenweber and Reinking placed those forms with very long macroconidia, is also included in *F. avenaceum.*

Fusarium avenaceum mutates readily in culture. Pionnotal mutants occur in which the aerial mycelium is absent and the surface of the agar is covered by a sheet of macroconidia in pionnotes, thus intensifying the color of the culture. Mycelial mutants occur when sporodochium production and color production is suppressed (see color plates). In most cases however, sufficient distinctive macroconidia are produced on CLA for positive identification.

We have not found isolates of *F. avenaceum* that produce polyphialides. Booth (4) placed *F. avenaceum, F. semitectum, F. camptoceras, F. sporotrichioides,* and *F. chlamydosporum* (=*F. fusarioides*) in the section Arthrosporiella based on the presence of polyphialides and microconidia. We have examined many isolates of *F. avenaceum* and have been unable to find polyphialides in any of these isolates. Early stages of branching can be mistaken for polyphialides (Fig. 20c, d). Nirenberg (48) reports the presence of polyphialides and microconidia in *F. arthrosporioides.* We have not found true polyphialides in this species and this report also may be early stages of branching mistaken for polyphialides (Fig. 20c, d).

Fusarium avenaceum is cosmopolitan in distribution.

Fusarium avenaceum has been reported to be toxigenic.

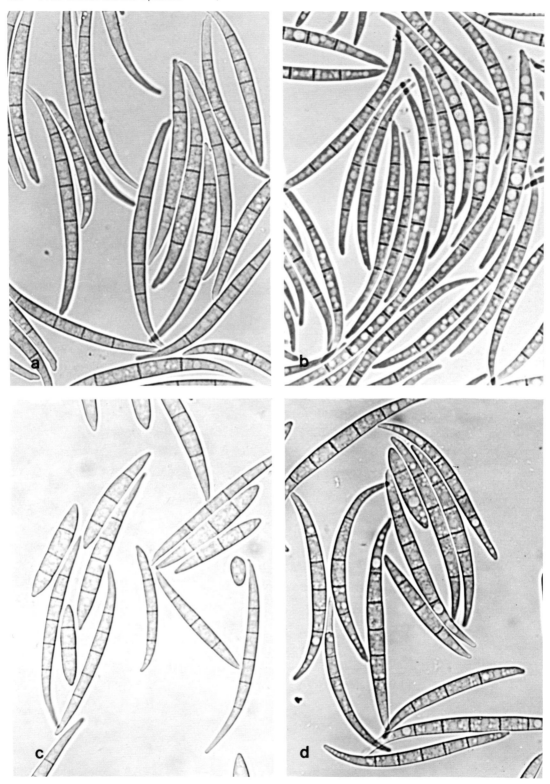

Figure 19. *Fusarium avenaceum:* a, b, macroconidia; c, d, macroconidia and microconidia (X1000).

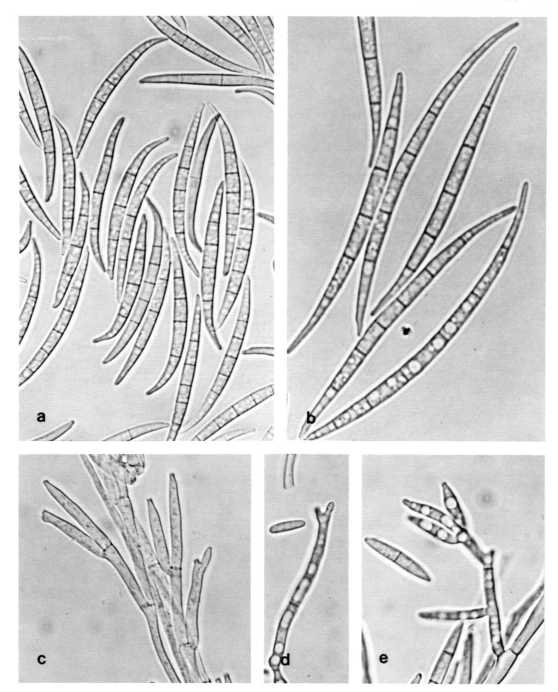

Figure 20. *Fusarium avenaceum:* a, b, macroconidia; c–e, conidiophores (unbranched and branched monophialides) (X1000).

6. SECTION ARTHROSPORIELLA

12. *Fusarium semitectum* Berk. & Rav. [W&R,G,B,J]

Syn.: *F. roseum* Lk. emend. Snyd. & Hans. pro parte [S&H]

F. roseum Lk. emend. Snyd. & Hans. var. *arthrosporioides* (Sherb.) Messiaen & Cassini pro parte [M&C]

Conidia (Figs. 21, 22a, b). Microconidia are scarce. Macroconidia may be of two types. Those borne in the aerial mycelium are spindle-shaped, straight to slightly curved, and the basal cell has a papilla but is not foot-shaped. The majority of the sickle-shaped macroconidia are borne in sporodochia, in those clones that produce them. These macroconidia are slightly curved, with a foot-shaped basal cell, and resemble macroconidia produced by some isolates of *F. equiseti, F. reticulatum,* or *F. heterosporum.*
Conidiophores (Fig. 22c,d,g). Unbranched and branched monophialides and polyphialides.
Chlamydospores (Fig. 22e,f). Present.
Perfect state. None known.
Colony morphology. On PDA, growth is rapid, with dense aerial mycelium that is tan to brown colored. Sporodochia may be present or absent and are light orange in color. The undersurface varies from tan to brown in color and darker brown flecks may also be present. It is rarely carmine red.

Most distinguishing characteristics. The presence of polyphialides in the aerial mycelium and the spindle-shaped macroconidia produced in the aerial mycelium.
Affinities. *Fusarium semitectum* resembles *F. equiseti* in colony morphology and color. The surface of the aerial mycelium is more powdery and the presence of the spindle-shaped macroconidia separates *F. semitectum* from *F. equiseti.*

Wollenweber and Reinking (76) considered *F. semitectum* as encompassing only those clones producing no sporodochia. We find this to be a variable phenomenon, particularly when cultures grown on PDA and CLA are compared. We place all those forms producing spindle-shaped macroconidia on polyphialides in the aerial mycelium in *F. semitectum.* When identifying such isolates, it is important to remember that since the typical sickle-shaped macroconidia are formed mainly in sporodochia and the spindle-shaped macroconidia are formed primarily in aerial mycelium, macroconidia from both sources should be examined.

Wollenweber and Reinking (76) recognized *F. semitectum* var. *majus* based on the production of larger spindle-shaped macroconidia; we include it in *F. semitectum.*

Fusarium semitectum can mutate to a white, featureless mycelial type or to a pionnotal type which has a wet, slimy, tan to orange color on the upper surface of PDA. In gross morphology these mutants may resemble mutants formed by *F. equiseti* on PDA. However, typical spindle-shaped macroconidia of *F. semitectum* still occur.

Fusarium semitectum is cosmopolitan.

Fusarium semitectum has been reported to be toxigenic.

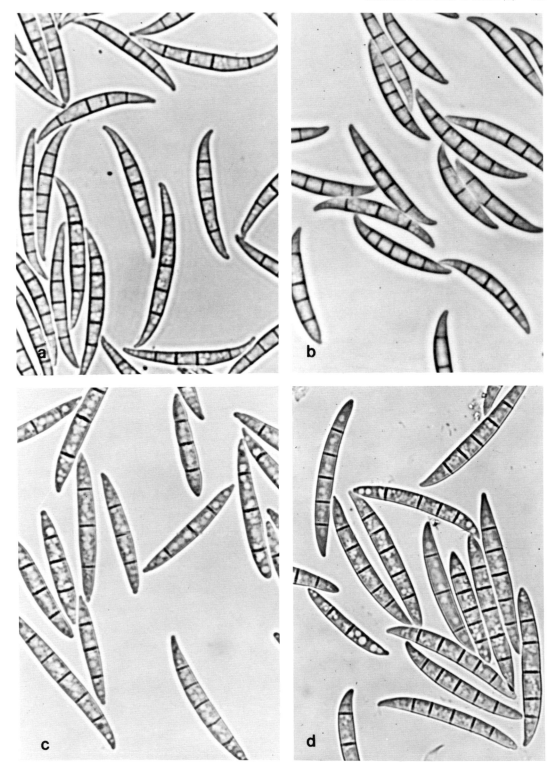

Figure 21. *Fusarium semitectum:* a, b, macroconidia formed in sporodochia; c, d, macroconidia formed in the aerial mycelium (X1000).

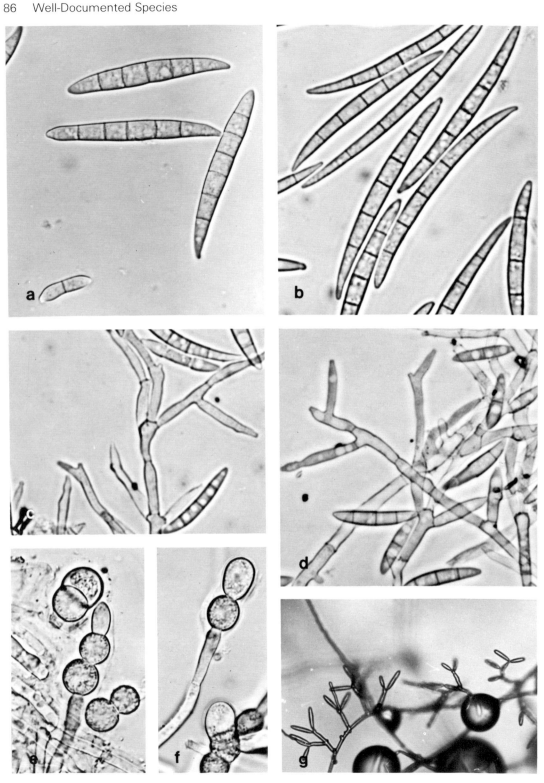

Figure 22. *Fusarium semitectum:* a, b, macroconidia formed in the aerial mycelium; c, d, polyphialides; e, f, chlamydospores; g, macroconidia borne on polyphialides (a–f, X1000; g X320).

6. SECTION ARTHROSPORIELLA
13. *Fusarium camptoceras* Wollenw. & Reinking [W&R,G,B,J]

Syn.: *F. roseum* Lk. emend. Snyd. & Hans. pro parte [S&H]

F. roseum Lk. emend. Snyd. & Hans. var. *arthrosporioides* (Sherb.) Messiaen & Cassini pro parte [M&C]

Conidia (Fig. 23a,b). Microconidia are scarce. Macroconidia are curved but otherwise resemble the spindle-shaped macroconidia of *F. semitectum.* The basal cell has a papilla but is not foot-shaped.
Conidiophores (Fig. 23c–e). Unbranched and branched monophialides and polyphialides.
Chlamydospores. Present.
Perfect State. None known.
Colony morphology. On PDA growth is rapid and the aerial mycelium is tan colored. Sporodochia are absent. The undersurface is tan to brown in color.

Most distinguishing characteristics. Polyphialides in the aerial mycelium, producing curved, spindle-shaped macroconidia of the type produced in *F. semitectum.*
Affinities. *Fusarium camptoceras* resembles *F. semitectum* in most respects except that the macroconidia are curved.

The characteristic curved macroconidia do not resemble those illustrated for *F. camptoceras* by Booth (4). We have not seen many isolates of this species.

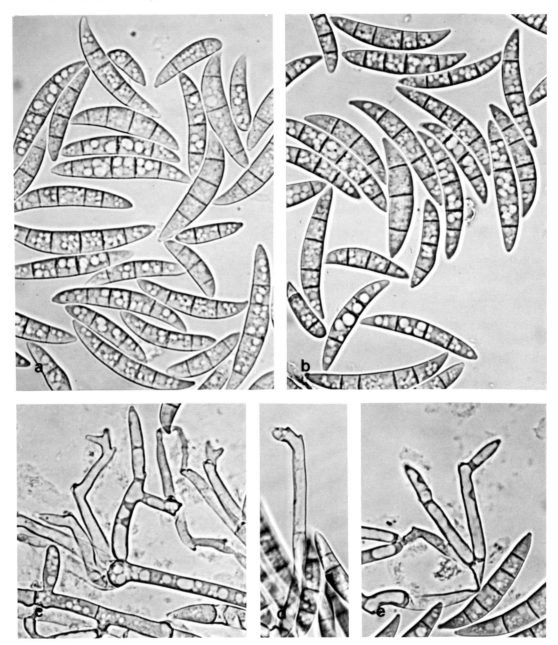

Figure 23. *Fusarium camptoceras:* a, b, macroconidia produced in the aerial mycelium; c–e, polyphialides (X1000).

7. SECTION GIBBOSUM
14. *Fusarium equiseti* **(Corda) Sacc.** sensu Gordon (24) [G,B,J]

Syn.: *F. equiseti* (Corda) Sacc. pro parte [W&R]

F. scirpi Lambotte & Fautr. var. *compactum* Wollenw. [W&R]

F. scirpi Lambotte & Fautr. var. *filiferum* (Preuss) Wollenw. [W&R]

F. roseum Lk. emend. Snyd. & Hans. 'Equiseti' pro parte [S&H]

F. roseum Lk. emend. Snyd. & Hans. 'Gibbosum' pro parte [S & H]

F. roseum Lk. emend. Snyd. & Hans. var. *gibbosum* (Wollenw.) Messiaen & Cassini pro parte [M&C]

Conidia (Figs. 24, 25, 26b). Microconidia may be present in the aerial mycelium and are oval to comma-shaped. The macroconidia are strongly septate, thick-walled, and sickle-shaped with a distinctive curvature. The ventral side is smoothly arched but the dorsal side is curved more abruptly. The apical cell is more or less elongated and emphasizes the curvature of the spore. The apical cell may be very long and whip-like (see *F. scirpi* var. *filiferum* below). The basal cell is distinctly foot-shaped, sometimes in an exaggerated manner.

Conidiophores (Fig. 26a). Unbranched and branched monophialides.

Chlamydospores (Fig. 26c–f). They are abundant with very prominent thick, roughened walls, and are produced in clumps or chains.

Perfect state. *Gibberella intricans* Wollenw. (5).

Colony morphology. On PDA, growth is rapid, with dense aerial mycelium that is white at first and becomes tan to brown as the culture ages. Orange sporodochia may appear as the culture ages. The under surface varies from tan to brown, with brown flecks, and sometimes may be carmine red (see *F. scirpi* var. *compactum* below).

Most distinguishing characteristics. The shape of the macroconidia, the extended apical cell, and the distinctive foot-shaped basal cell. On PDA the cultures look like *F. semitectum,* but no polyphialides are present. There is abundant development of chlamydospores in some clones.

Affinities. *Fusarium equiseti* resembles *F. semitectum* in colony morphology and color. However, the shape of the macroconidia produced in the aerial mycelium and sporodochia on CLA is distinctive.

Wollenweber and Reinking (76) separated the species *F. equiseti* and *F. scirpi* on the basis of differences in spore curvature. Within *F. equiseti* they maintained the var. *bullatum,* based on further differences in conidial shape; within *F. scirpi* they maintained the var. *caudatum,* also based on variations in shape of the conidia. We consider these differences to be within the range of variation encompassed by *F. equiseti.* *Fusarium scirpi* var. *compactum* (Figs. 27, 28) was separated by Wollenweber and Reinking (76) from the base species because the macroconidia are shorter, fatter, and more compact and because of the carmine red undersurface. We have seen many such isolates and they appear to be a distinct group. We are provisionally placing them under *F. equiseti* in this publication.

Fusarium scirpi var. *filiferum* (Fig. 25c) is separated from the base species by Wollenweber and Reinking (76) because of macroconidia with whip-like apical cells and exaggerated foot-shaped basal cells produced in sporodochia. No carmine red color is produced in the undersurface. We have seen many cultures that fit this description. However, in many other cultures macroconidia are produced that grade from the *F. scirpi* var. *filiferum* type to the *F. equiseti* type without any sharp distinction. Because of the lack of a clear-cut demarcation, we are provisionally placing these isolates, considered to be *F. scirpi* var. *filiferum,* within *F. equiseti.*

Fusarium equiseti frequently mutates in culture to the pionnotal form, in which aerial mycelium is absent and the surface is covered by a sheet of conidia on pionnotes. The cultures have a wet, yeast-like appearance and are bright orange to tan. However, the characteristic shape of the macroconidium usually does not vary.

It is important to note that in sporodochial forms, the characteristically shaped macroconidia are formed in the sporodochia which develop well on CLA. The aerial mycelium produces a large and bewildering array of spore types, including macroconidia of various shapes and sizes and microconidia of various shapes and sizes, which tend to make proper diagnosis more difficult.

Fusarium equiseti is cosmopolitan.

Fusarium equiseti has been reported to be toxigenic.

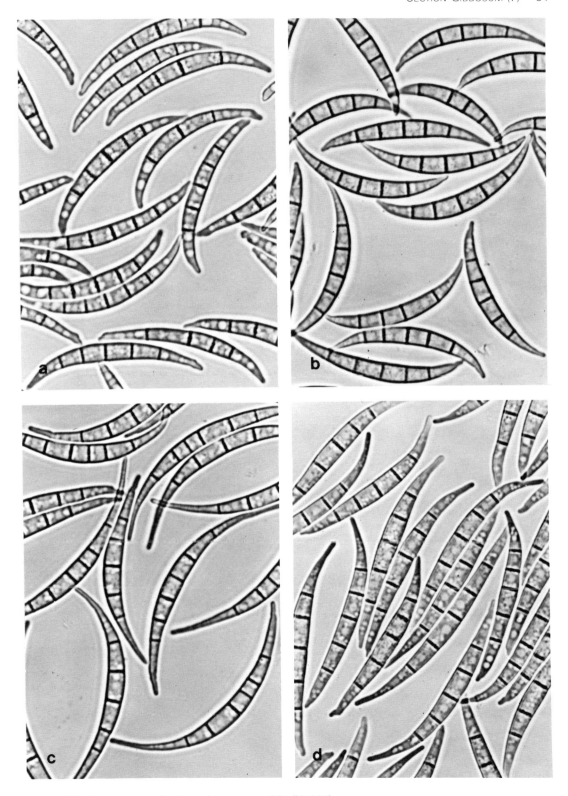

Figure 24. *Fusarium equiseti:* a–d, macroconidia (X1000).

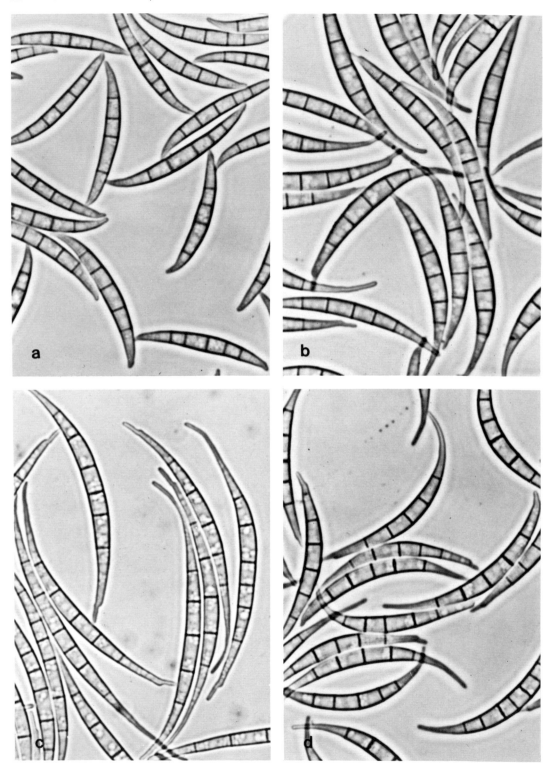

Figure 25. *Fusarium equiseti:* a, b, d, macroconidia; c, macroconidia of the *F. scirpi* var. *filiferum* type (X1000).

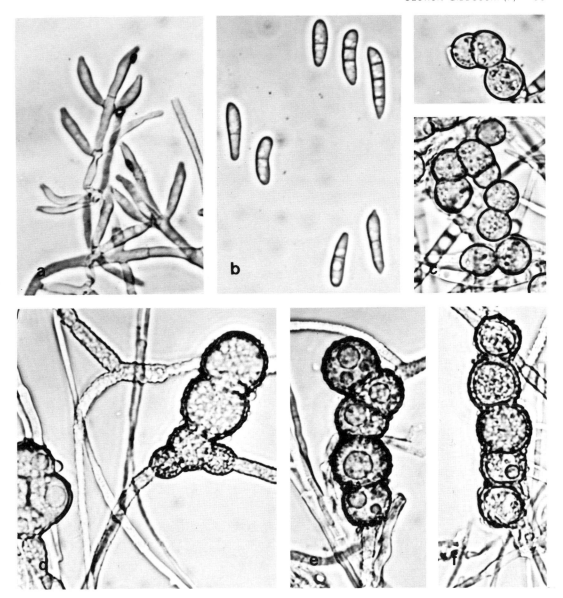

Figure 26. *Fusarium equiseti:* a, microconidiophores (monophialides); b, microconidia; c–f, chlamydospores (X1000).

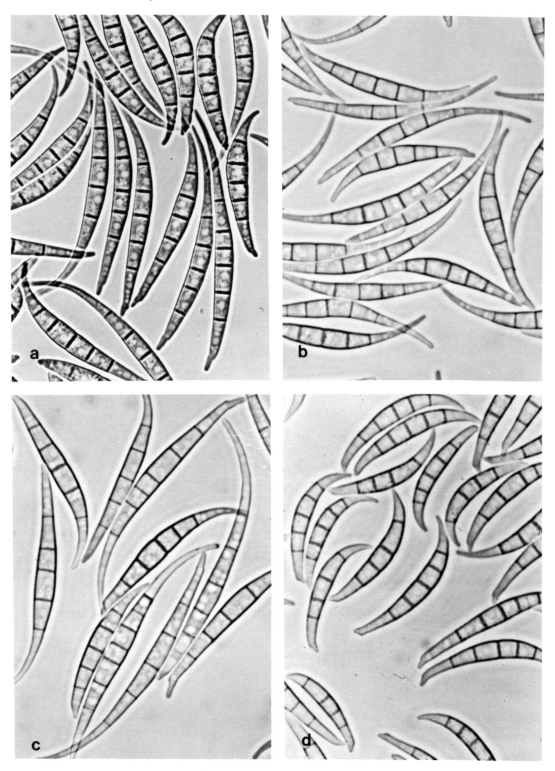

Figure 27. *Fusarium equiseti* (*F. scirpi* var. *compactum*): a–d, macroconidia produced in sporodochia (X1000).

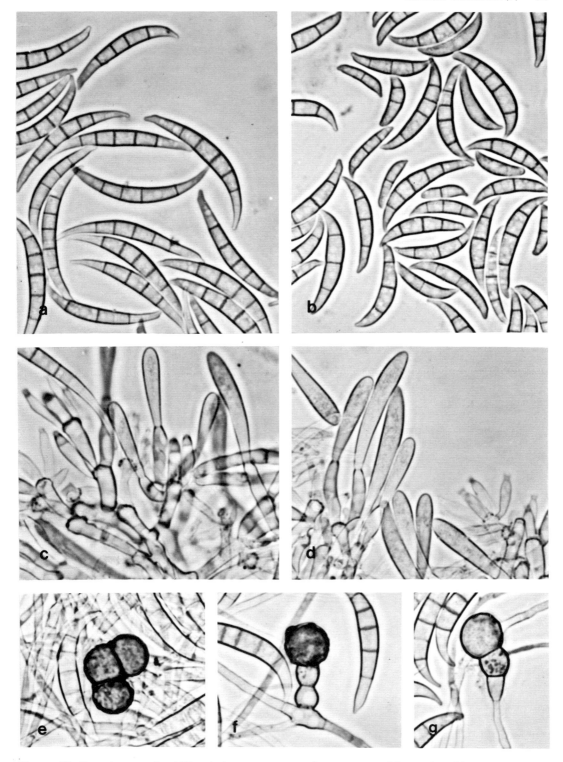

Figure 28. *Fusarium equiseti* (*F. scirpi* var. *compactum*): a, macroconidia produced in sporodochia; b, macroconidia produced in the aerial mycelium; c, d, macroconidiophores (monophialides); e–g, chlamydospores (X1000).

7. SECTION GIBBOSUM

15. *Fusarium scirpi* Lambotte & Fautr.

Syn.: *F. scirpi* Lambotte & Fautr. pro parte [W&R]
F. equiseti (Corda) Sacc. pro parte [G,B,J]
F. roseum Lk. emend. Snyd. & Hans. 'Equiseti' pro parte [S&H]
F. roseum Lk. emend. Snyd. & Hans. 'Gibbosum' pro parte [S&H]
F. roseum Lk. emend. Snyd. & Hans. var. *gibbosum* (Wollenw.) Messiaen & Cassini pro parte [M&C]

Conidia (Figs. 29, 30a,b). Microconidia are present, ellipsoidal to club-shaped, and 0–3 septate. Macroconidia resemble those formed by *F. equiseti* and are distinctly septate, thick-walled, sickle-shaped, with a distinctive curvature; the ventral side is smoothly arched but the dorsal side is more abruptly curved. The septa in the macroconidium are often closer together in the middle of the conidium than at the ends. The apical cell is more or less elongated and emphasizes the curvature of the spore. The basal cell is distinctly foot-shaped.

Conidiophores (Fig. 30c–e). Unbranched and branched monophialides and characteristic, short, truncate, often cross-shaped polyphialides bearing microconidia.

Chlamydospores (Fig. 30f,g). Present, borne in clumps or in chains.

Perfect state. None known.

Colony morphology. As in *F. equiseti,* the growth on PDA is rapid, with dense aerial mycelium at first white and later turning to tan or brown. Orange sporodochia may appear as the culture ages. The undersurface varies from tan to brown and may have brown flecks.

Most distinguishing characteristics. A fungus similar to *F. equiseti* but with abundant microconidia formed on short, often cross-shaped polyphialides. On close inspection the culture appears to have a powdery look resulting from the presence of the polyphialides bearing microconidia.

Affinities. *Fusarium scirpi* is similar in colony morphology on PDA and in the shape of macroconidia on CLA to *F. equiseti,* but microconidia are more abundant and polyphialides are present in *F. scirpi.* Some cultures of *F. semitectum* are similar to cultures of *F. scirpi* on PDA but the shape of the polyphialides and the presence of microconidia, especially on CLA, separate *F. scirpi* from *F. semitectum.*

We are reexamining the entire concept of *F. scirpi* (8) as used by Wollenweber and Reinking (76). In his *Fusaria autographice Delineata* (74), Wollenweber illustrates polyphialides (Nos. 198, 595) similar to those shown in Fig. 30c–e and microconidia (Nos. 198, 199, 200, 201, 595) similar to those shown in Fig. 30a–b. These drawings are all labeled *F. chenopodinum,* a species which Wollenweber and Reinking (76) included later as a synonym of *F. scirpi.* Further studies are in progress to determine the limits of *F. scirpi* as used by Wollenweber and Reinking (76).

It is important to note that the microconidia and the polyphialides are in the aerial mycelium. Sampling of the sporodochia on CLA only is not adequate to make a correct identification.

Fusarium scirpi is relatively stable in culture.

Fusarium scirpi has been found in warm, semi-arid areas of New South Wales, Victoria, and Queensland, Australia, in the Cape Province of the Republic of South Africa, and in the Butterworth area of the Republic of Transkei.

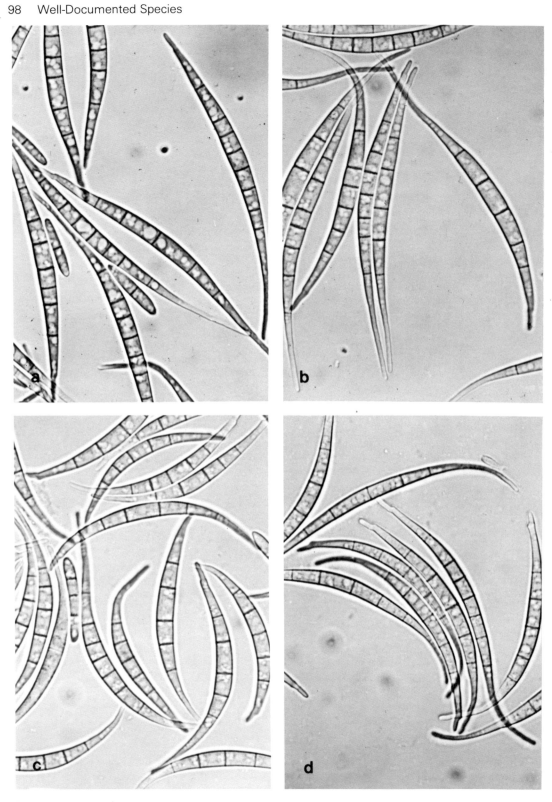

Figure 29. *Fusarium scirpi:* a–d, macroconidia produced in sporodochia (X1000).

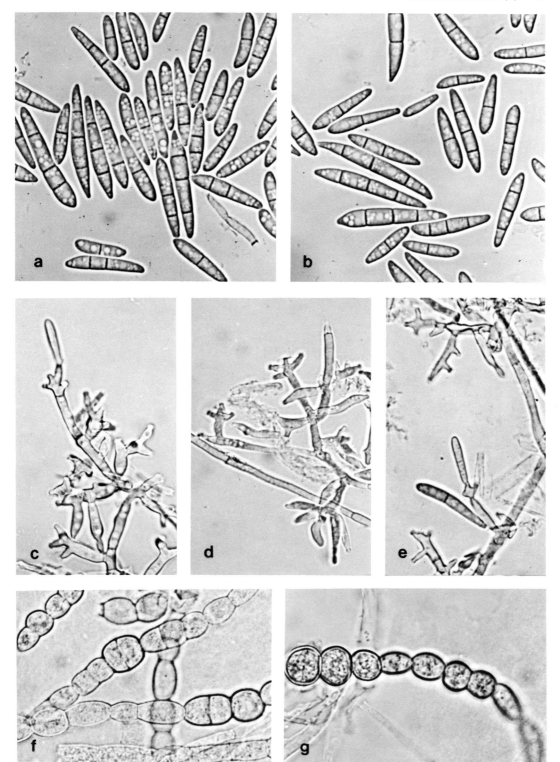

Figure 30. *Fusarium scirpi:* a, b, microconidia produced in the aerial mycelium; c–e, polyphialides; f, g, chlamydospores (X1000).

7. SECTION GIBBOSUM

16. *Fusarium acuminatum* Ell. & Ev. sensu Gordon (24) [G,B]

Syn.: *F. scirpi* Lambotte & Fautr. var. *acuminatum* (Ell. & Ev.) Wollenw. [W&R]
F. equiseti (Corda) Sacc. var. *acuminatum* (Ell. & Ev.) Bilai [J]
F. roseum Lk. emend. Snyd. & Hans. 'Acuminatum' [S&H]
F. roseum Lk. emend. Snyd. & Hans. var. *gibbosum* (Wollenw.) Messiaen
& Cassini pro parte [M&C]

Conidia (Figs. 31, 32, 33d). Microconidia generally are sparse. Macroconidia are thin and strongly curved into a sickle-shape, with the widest point one-third of the distance from the base. The apical cell is elongated as in *F. equiseti,* emphasizing the curvature of the spore. The basal cell is distinctly foot-shaped.
Conidiophores (Fig. 33a–c). Unbranched and branched monophialides.
Chlamydospores (Fig. 33e,f). They are present and are formed singly, in chains or in clumps. They may form slowly in culture in some clones.
Perfect state. *Gibberella acuminata* Wollenw. (5).
Colony morphology. On PDA growth is rapid, with dense white aerial mycelium, and orange sporodochia may develop as the culture ages. The undersurface is usually carmine red but may also be tan to brown.

Most distinguishing characteristics. The shape of the macroconidia and the presence of chlamydospores.
Affinities. *Fusarium acuminatum* generally exhibits the rapid, profuse mycelial growth and carmine red undersurface on PDA that are characteristic of some isolates in the sections Sporotrichiella, Arthrosporiella, Roseum, and Discolor.

One of the most difficult distinctions to make is the separation of *F. avenaceum* and *F. acuminatum* because the shapes of the macroconidia of these two species grade into one another (see *F. avenaceum,* Figs. 19, 20). A major distinction is chlamydospores, which are formed in *F. acuminatum* but not in *F. avenaceum.* This process can be very slow in some clones of *F. acuminatum.*

Fusarium acuminatum is relatively stable in culture but variants may occur.

Fusarium acuminatum is cosmopolitan.

Fusarium acuminatum has been reported to be toxigenic.

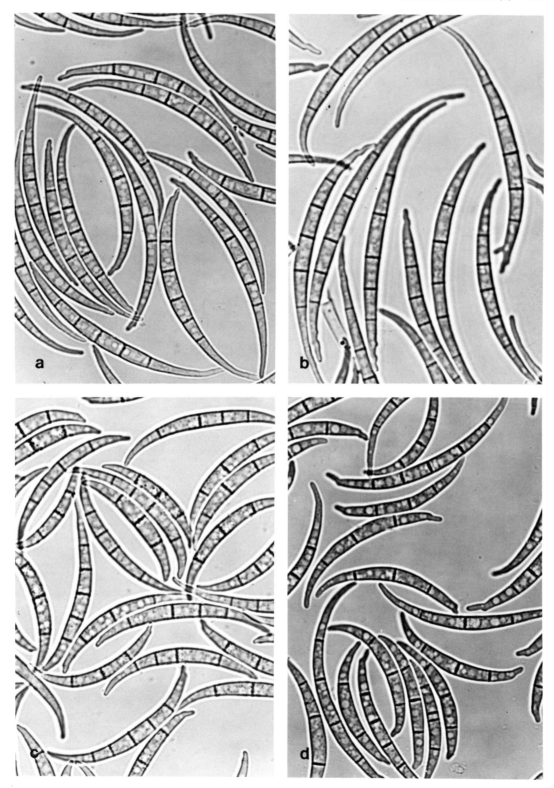

Figure 31. *Fusarium acuminatum:* a–d, macroconidia produced in sporodochia (X1000).

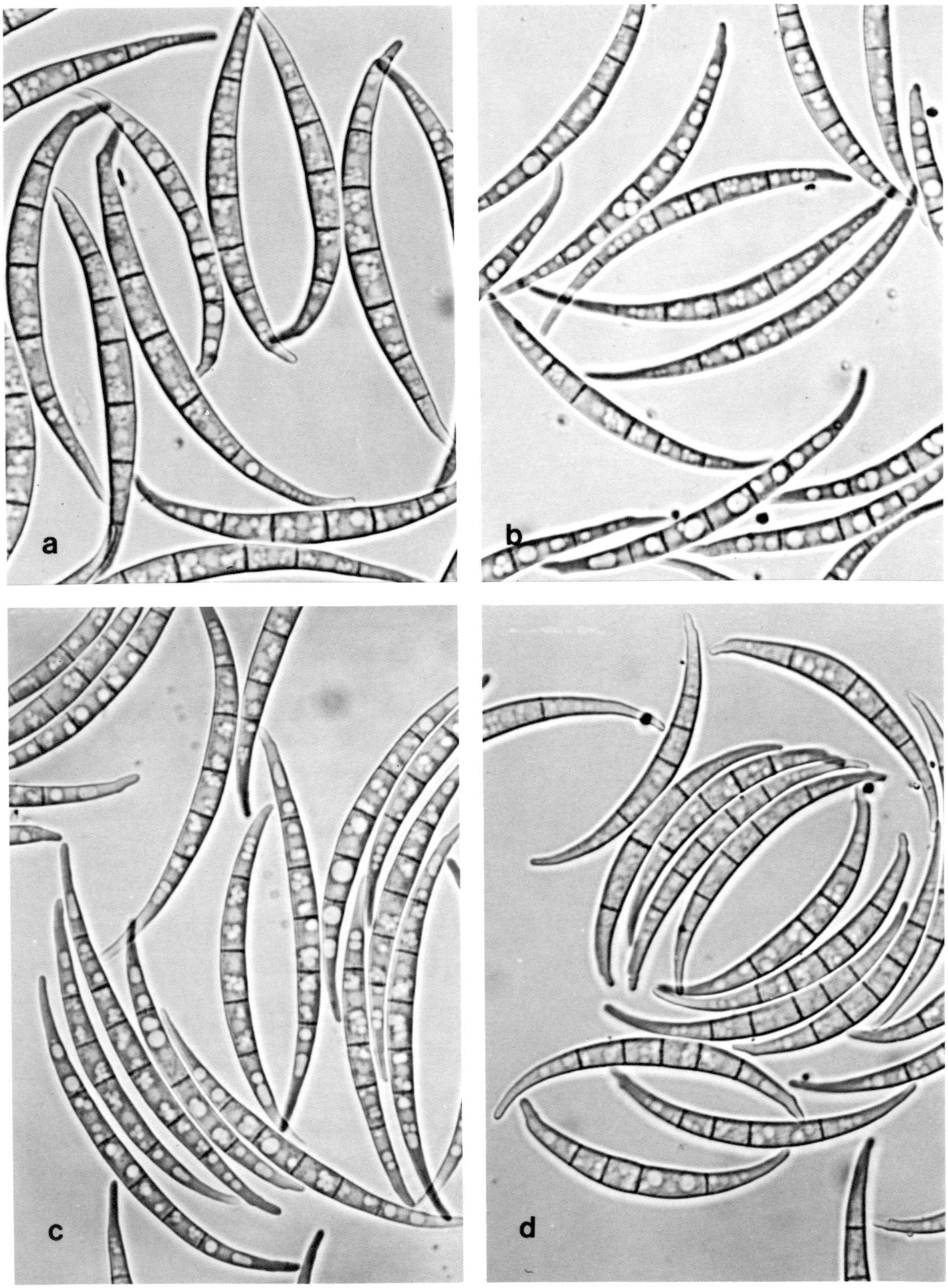

Figure 32. *Fusarium acuminatum:* a–d, macroconidia produced in sporodochia (X1000).

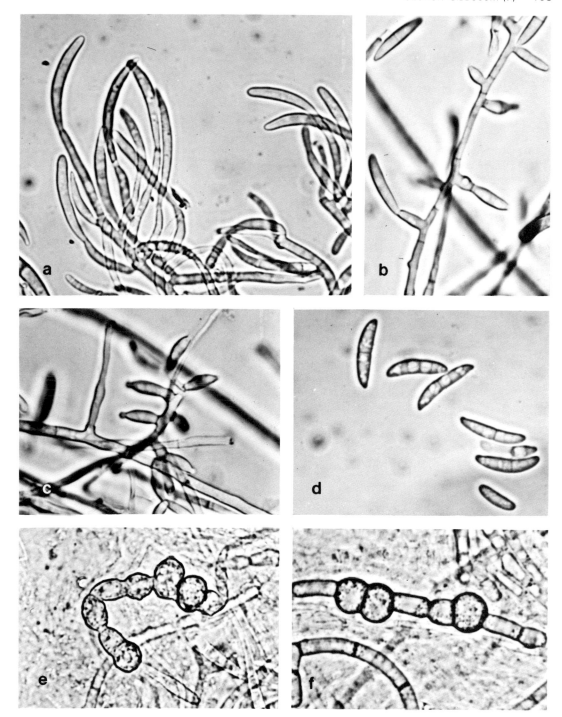

Figure 33. *Fusarium acuminatum:* a, macroconidiophores (monophialides); b, c, microconidiophores (monophialides); d, microconidia produced in the aerial mycelium; e, f, chlamydospores (X1000).

7. SECTION GIBBOSUM
17. *Fusarium longipes* Wollenw. & Reinking sensu Gordon (24) [G]

Syn.: *F. scirpi* Lambotte & Fautr. var. *longipes* (Wollenw. & Reinking) Wollenw.
[W&R]

F. equiseti (Corda) Sacc. pro parte [B]

F. equiseti (Corda) Sacc. var. *longipes* (Wollenw. & Reinking) Joffe [J]

F. roseum Lk. emend. Snyd. & Hans. pro parte [S&H,M&C]

Conidia (Figs. 34a,b, 35). Microconidia are generally absent. Macroconidia are thin, very long, curved in a sickle-shape similar to that of *F. equiseti,* where the dorsal surface has a sharper bend than the ventral surface. The apical cell is often elongated and whip-like and may be curved. The basal cell is distinctly foot-shaped, often in an exaggerated manner.
Conidiophores (Fig. 34c). Unbranched and branched monophialides.
Chlamydospores (Fig. 34d,e). They are present, and are generally single.
Perfect state. None known.
Colony morphology. On PDA growth is rapid, with dense white aerial mycelium often tinged with carmine red. Prominent tan to orange sporodochia occur as the culture ages. The undersurface is tan to carmine red.

Most distinguishing characteristics. The shape of the macroconidia—often with a whip-like apical cell and an extended foot-shaped basal cell which is very distinctive. The carmine red color of the undersurface of the colony on PDA.
Affinities. *Fusarium longipes* generally exhibits the rapid, profuse mycelial growth and carmine red undersurface on PDA that are characteristic of some isolates in the sections Sporotricheilla, Arthrosporiella, Roseum, and Discolor.

Wollenweber and Reinking (76) distinguished *F. scirpi* var. *longipes* from *F. scirpi* var. *filiferum* by the greater number of septations in the conidia and the absence of carmine color on the undersurface of the latter. We have included *F. scirpi* var. *filiferum* under *F. equiseti.* However, we have retained *F. longipes* as a distinct species because the shape of the macroconidia allows for ready separation from other members of the section Gibbosum.

Fusarium longipes is unstable in culture, rapidly mutating to the pionnotal form in which the aerial mycelium is absent and the colony surface is covered by macroconidia on pionnotes. The cultures then are carmine red on both surfaces.

Fusarium longipes is most abundant in wet tropical areas but also occurs in subtropical areas.

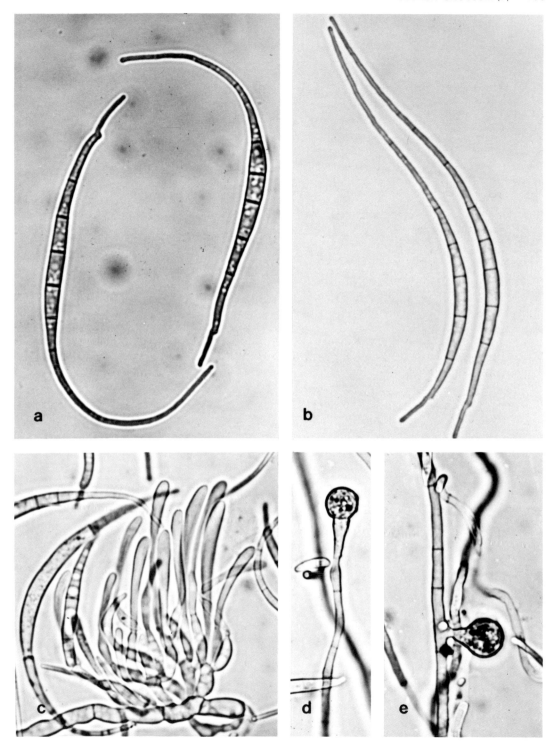

Figure 34. *Fusarium longipes:* a, b, macroconidia produced in sporodochia; c, macroconidiophores (monophialides); d, e, chlamydospores (X1000).

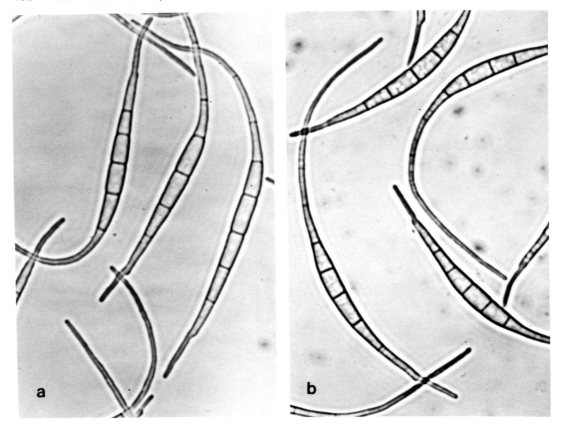

Figure 35. *Fusarium longipes:* a, b, macroconidia produced in sporodochia (X1000).

8. SECTION DISCOLOR
18. *Fusarium heterosporum* Nees [W&R,G,J]

Syn.: *F. heterosporum* Nees pro parte [B]
F. roseum Lk. emend. Snyd. & Hans. pro parte [S&H, M&C]

Conidia (Fig. 36a,b). Microconidia are absent. Macroconidia are sickle-shaped, narrowed at both ends, and pedicellate.
Conidiophores (Fig. 36c,d). Unbranched and branched monophialides.
Chlamydospores. They are present and are formed in chains.
Perfect state. *Gibberella gordonii* Booth (5).
Colony morphology. On PDA growth is rapid, with dense white to pink aerial mycelium. Orange sporodochia develop as the culture ages. The undersurface is light orange to tan in color.

Most distinguishing characteristics. The colony appearance on PDA and the bright orange sporodochia formed in fluffy white to pink aerial mycelium, along with the spore morphology on CLA.
Affinities. *Fusarium heterosporum* resembles *F. graminum* when grown on PDA, but the macroconidia formed on CLA are quite different and chlamydospores are present.

Wollenweber and Reinking (76) mention a variety, *F. heterosporum* var. *congoense,* which is differentiated by larger conidia and the carmine red undersurface. We have not seen this type and we have rarely encountered *Fusarium heterosporum.* Booth (4) considers *F. reticulatum* and *F. graminum* as synonyms of *F. heterosporum,* but we have retained both as distinct species.

Fusarium heterosporum appears to be stable in culture but can mutate to a mycelial type, resulting in a colony that is white and featureless.

Fusarium heterosporum has been reported to be toxigenic.

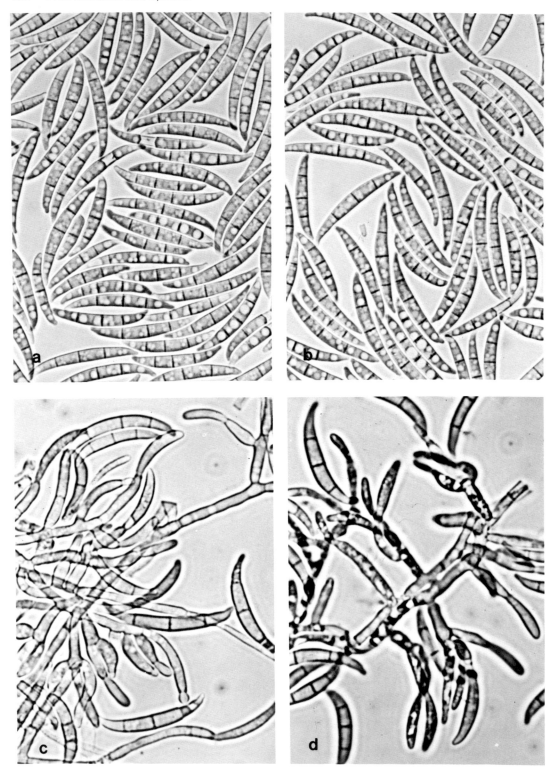

Figure 36. *Fusarium heterosporum:* a, b, macroconidia; c, d, conidiophores (monophialides) (X1000).

8. SECTION DISCOLOR
19. *Fusarium reticulatum* Mont. [W&R,G]

Syn.: *F. heterosporum* Nees pro parte [B]
F. roseum Lk. emend. Snyd. & Hans. pro parte [S&H,M&C]
[Species not mentioned by Joffe]

Conidia (Fig. 37a–c). Microconidia are absent. Macroconidia are sickle-shaped, narrowed at both ends, generally 3 septate but may be 5 septate, and pedicellate.
Conidiophores (Fig. 37d). Unbranched and branched monophialides.
Chlamydospores (Fig. 37e). They are present and are formed in chains or in clumps.
Perfect state. *Gibberella cyanea* (Sollm.) Wollenw. (76).
Colony morphology. On PDA growth is slow to moderate, with dense white aerial mycelium often tinged with carmine red. Orange sporodochia may develop as the culture ages. The undersurface is usually carmine red.

Most distinguishing characteristics. The shape of the macroconidia (Fig. 37a–c). This fungus produces the most slender (generally 3 septate) macroconidia of the section Discolor, but its growth rate is slower than the other species.
Affinities. *Fusarium reticulatum* has a slow to moderate growth rate on PDA and the carmine red color of the undersurface of cultures grown on PDA is characteristic of the section Discolor. Except for growth rate, cultural morphology is similar to *F. culmorum, F. crookwellense,* and *F. graminearum,* and the carmine red clones of *F. sambucinum* when grown in PDA. The morphology of the macroconidia produced on CLA is quite different from the species mentioned above in that the macroconidia are smaller and more slender, like those produced by *F. heterosporum.* The macroconidia of *F. reticulatum* also resemble those of the *Fusarium* species in Wollenweber and Reinking's (76) section Sporotrichiella. Members of the section Sporotrichiella are different in that they produce abundant microconidia and have a more rapid growth rate. Confusion may occur if only sporodochia on CLA plates are examined.

Wollenweber and Reinking (76) differentiate a form 1 and the variety *F. reticulatum* var *negundinis* based on variations in colony color and conidial size. We group both the form and the variety under the one species, *F. reticulatum.*

Fusarium reticulatum appears to be relatively stable in culture.

Fusarium reticulatum is cosmopolitan.

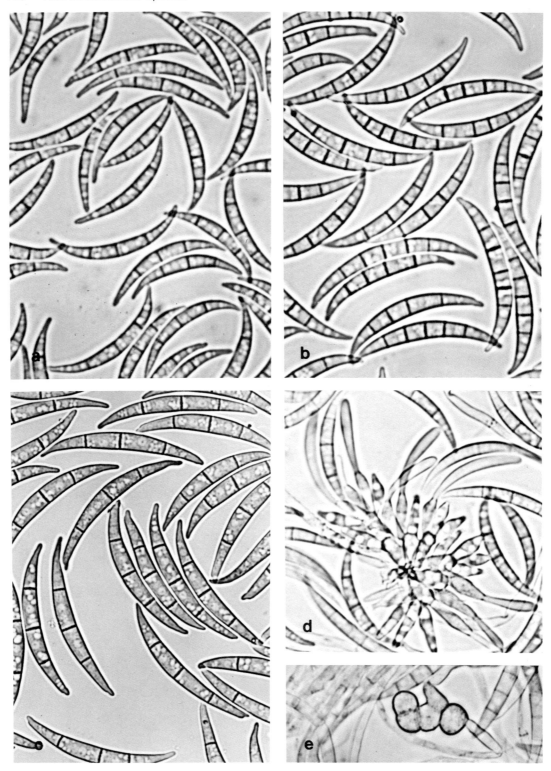

Figure 37. *Fusarium reticulatum:* a–c, macroconidia produced in sporodochia; d, conidiophores (monophialides); e, chlamydospores (X1000).

8. SECTION DISCOLOR
20. *Fusarium sambucinum* Fuckel [W&R,G,B,J]

Syn.: *F. trichothecioides* Wollenw. [W&R,G,B]

F. bactridioides Wollenw. [W&R,G]

F. sulphureum Schlecht. [G,B]

F. sambucinum Fuckel var. *trichothecioides* (Wollenw.) Bilai [J]

F. roseum Lk. emend. Snyd. & Hans. 'Sambucinum' [S&H]

F. roseum Lk. emend. Snyd. & Hans. var. *sambucinum* (Fuckel) Snyd. & Hans. [M&C]

Conidia (Figs. 38, 39a,b). Microconidia are generally absent (see comments below on *F. bactridioides*). Macroconidia are short and stout, distinctly septate, with thick walls and moderate to strongly curved ventral and dorsal surfaces. The basal cell varies from distinctly foot-shaped to notched to rounded. The apical cell is constricted into a snout, often with a small papilla, and is sometimes strongly curved.

Conidiophores (Fig. 39c,d). Unbranched and branched monophialides.

Chlamydospores (Fig. 39e). Generally they are formed abundantly and quickly (less than 1 month) and are single, in chains, or in clumps.

Perfect state. *Gibberella pulicaris* (Fr.) Sacc. (5).

Colony morphology. On PDA growth is rapid, with or without dense aerial mycelium. When aerial mycelium is present it is white, tan, or pink to reddish brown in color. Dark blue sclerotia may be present. Cream to tan to orange sporodochia or pionnotes may be present or absent. The undersurface may be of various colors; it is frequently carmine red, but also may be tan to brown.

Most distinguishing characteristics. The shape of the macroconidia on CLA (Figs. 38, 39a,b). The macroconidia resemble those of *F. culmorum* but they are thinner and the constriction and/or curvature of the apical cell is more pronounced.

Affinities. On PDA, *F. sambucinum* often has the typical rapid growth and carmine red undersurface characteristic of the species in the section Discolor. The morphology of the macroconidia show strong affinities to other species of the section Discolor, especially when compared with *F. culmorum*.

Fusarium bactridioides Wollenw. produces macroconidia similar to *F. sambucinum* and also produces abundant 0–3 septate microconidia (Fig. 39a,b). We have seen many examples of this type but have elected to place them provisionally within *F. sambucinum* because we have not found a clear demarcation line between these two species. The presence or absence of microconidia is not clear cut and there is a gradation of forms between the two extremes.

In addition, Wollenweber and Reinking (76) recognize *F. trichothecioides*, *F. sambucinum* var. *minus*, and five forms of *F. sambucinum*. The formae and varieties are separated mainly on the basis of colony colors. Gerlach (21) and Booth (4) recognize *F. sambucinum* var. *coeruleum* (perfect state *G. pulicaris* var. *minor*) and *F. sulphureum* (perfect state *G. cyanogena*). We have grouped all of these under *F. sambucinum*. This makes for a large, and from the point of view of cultural appearance on PDA, a rather varied group. This variation is increased by the presence of pionnotal and mycelial mutants which may also occur in nature, especially on a substrate such as potato tubers. Nevertheless, the unifying feature of macroconidial shape makes all these forms relatively easy to identify as *F. sambucinum*.

Fusarium sambucinum is cosmopolitan.

Fusarium sambucinum has been reported to be toxigenic.

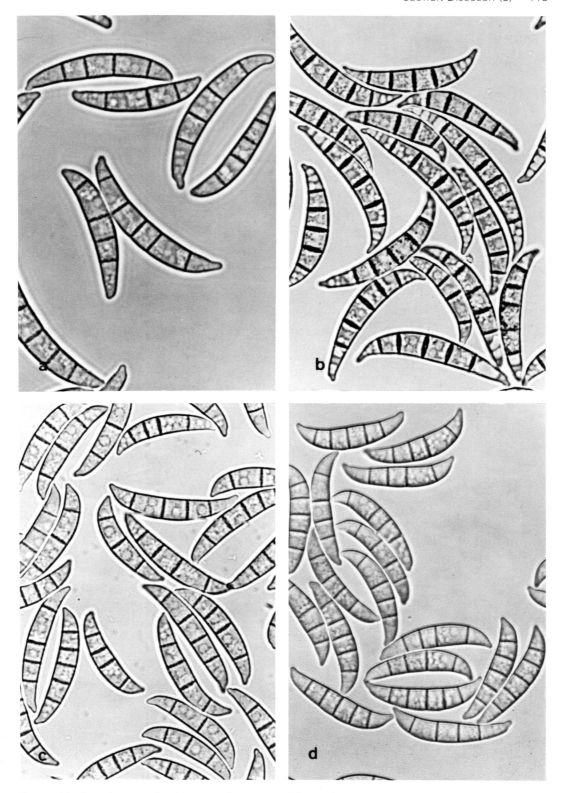

Figure 38. *Fusarium sambucinum:* a–d, macroconidia produced in sporodochia (X1000).

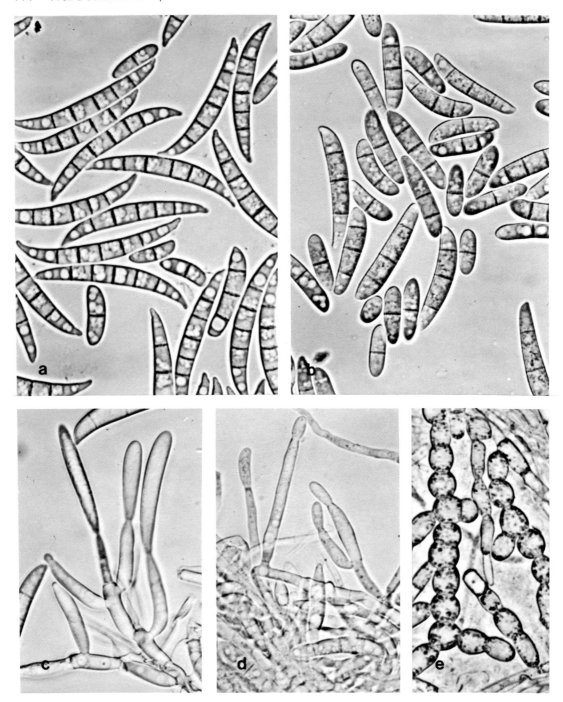

Figure 39. *Fusarium sambucinum:* a, b, macroconidia and microconidia produced in the aerial mycelium; c, macroconidiophores (monophialides); d, microconidiophores (monophialides); e, chlamydospores (X1000).

8. SECTION DISCOLOR

21. *Fusarium culmorum* (W. G. Smith) Sacc. [W&R,G,B,J]

Syn.: *F. roseum* Lk. emend. Snyd. & Hans. 'Culmorum' [S&H]

F. roseum Lk. emend. Snyd. & Hans. var. *culmorum* (Schwabe) (sic) Snyd. & Hans. [M&C]

Conidia (Fig. 40). Microconidia are absent. Macroconidia are stout, distinctly septate, thick-walled, and have curved ventral and dorsal surfaces. The basal cell varies from distinctly foot-shaped to slightly notched.

Conidiophores. Unbranched and branched monophialides.

Chlamydospores (Fig. 41). They generally form abundantly and quickly; they may occur singly, in chains, or in clumps.

Perfect state. None known.

Colony morphology. On PDA growth is rapid, with dense aerial mycelium, generally white but often yellow to tan toward the base of the slant. Orange to red-brown sporodochia appear as the culture ages. The undersurface is carmine red.

Most distinguishing characteristics. The shape of the macroconidia (Fig. 40). This fungus produces the most distinct short, stout macroconidia of the section Discolor.

Affinities. *Fusarium culmorum* has the typical rapid, profuse mycelial growth and carmine red undersurface on PDA characteristic of species in the section Discolor. On PDA, *F. culmorum* resembles *F. graminearum, F. crookwellense,* and certain clones of *F. sambucinum,* all in the section Discolor. On PDA the abundant sporulation around the point of inoculation separates it from *F. graminearum* but not from *F. crookwellense.*

Fusarium culmorum is a relatively stable fungus in culture, but mutants may occur. The mycelial mutant lacks color in the aerial mycelium, sporodochial formation is suppressed, and the carmine red pigment may be decreased. In pionnotal mutants the aerial mycelium is not present, and the yeast-like surface of the colony on PDA is a sheet of macroconidia on pionnotes. Both surfaces of the colony are then carmine. Wollenweber and Reinking (76) separated the variety, *F. culmorum* var. *cereale,* differentiated by longer and more slender macroconidia. We consider this distinction impractical.

Fusarium culmorum is cosmopolitan.

Fusarium culmorum has been reported to be toxigenic.

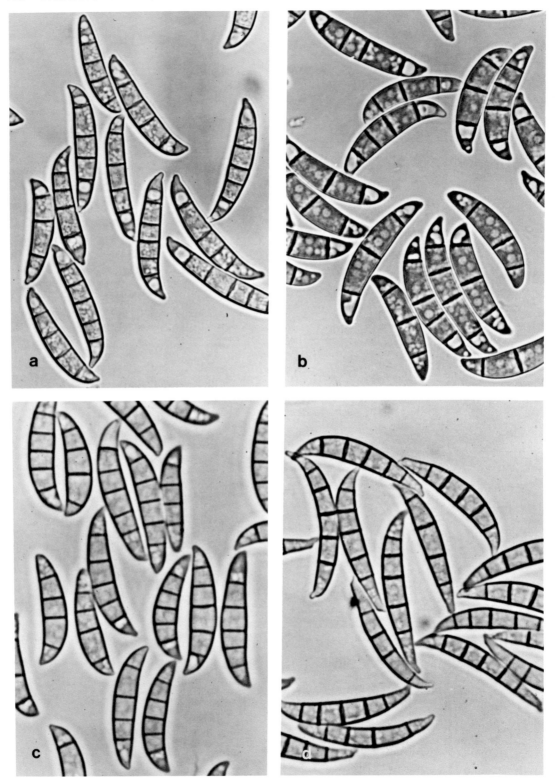

Figure 40. *Fusarium culmorum:* a–d, macroconidia produced in sporodochia (X1000).

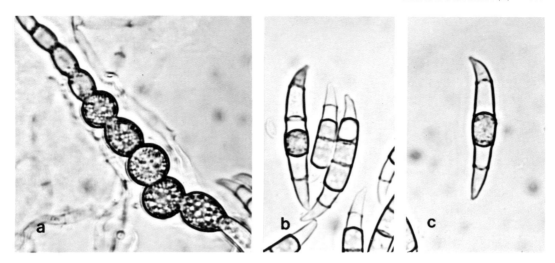

Figure 41. *Fusarium culmorum:* a–c, chlamydospores (X1000).

8. SECTION DISCOLOR
22. *Fusarium graminearum* Schwabe [W&R,G,B,J]

Syn.: *F. roseum* Lk. emend. Snyd. & Hans. 'Graminearum' [S&H]

F. roseum Lk. emend. Snyd. & Hans. var. *graminearum* (Schwabe) Snyd. & Hans. [M&C]

Conidia (Fig. 42). Microconidia are absent. Macroconidia are distinctly septate, thick-walled, straight to moderately sickle-shaped, unequally curved with the ventral surface almost straight and a smoothly arched dorsal surface. The basal cell is distinctly foot-shaped. The apical cell is cone-shaped or constricted as a snout.
Conidiophores (Fig. 43a). Unbranched and branched monophialides.
Chlamydospores (Fig. 43b,c). They are generally very slow to form in culture; when they do occur, they most often form in the macroconidia but may also form in the mycelium.
Perfect state. *Gibberella zeae* (Schw.) Petch (5).
Colony morphology. On PDA, growth is rapid, with dense aerial mycelium that may almost fill the tube and is frequently yellow to tan with the margins white to carmine red. Red-brown to orange sporodochia, if present, are sparse, often appearing only when the cultures are more than 30 days old. The undersurface is usually carmine red.

Most distinguishing characteristics. The shape of the macroconidia on CLA. This fungus produces the most cylindrical (dorsal and ventral surfaces parallel) macroconidia of any species of the section Discolor.
Affinities. *Fusarium graminearum* has the typical rapid, profuse mycelial growth and carmine red undersurface on PDA characteristic of most species in the section Discolor. On PDA cultures resemble *F. culmorum, F. crookwellense,* and certain clones of *F. sambucinum.* The sparse formation of sporodochia on PDA and the formation of perithecia (certain homothallic strains only) on CLA help to distinguish it from these species.

Fusarium graminearum exists as two distinct populations (10): group 1 prevalent in Australia and California, and group 2 prevalent in the eastern U.S.A. and Europe. On PDA, the aerial mycelium of group 1 is white to yellow to tan and is extremely abundant, often filling the tube. In group 2 the aerial mycelium is not quite so abundant and its color is more tan, grading to pink at the base of the slant. The two groups cannot be differentiated reliably on the basis of the conidia or conidiophores. The most distinctive difference between the two is the formation of perithecia. In group 2 the isolates are homothallic and perithecia appear on CLA as early as 4 days after single sporing. Group 1 isolates apparently are heterothallic and the perfect state has not been produced from a single macroconidium on CLA. Group 1 isolates cause a foot and crown rot of wheat. Group 2 isolates cause head scab of wheat and also attack corn and carnation.

Fusarium graminearum is a relatively stable fungus in culture, but mutants do occur. The mycelial mutant lacks color in the aerial mycelium, no sporodochia are formed, and the carmine red color of the undersurface may be decreased. In pionnotal mutants the aerial mycelium is absent and the colony surface is covered by a sheet of macroconidia in pionnotes. Both surfaces are then carmine red.

Fusarium graminearum is cosmopolitan.

Fusarium graminearum has been reported to be toxigenic.

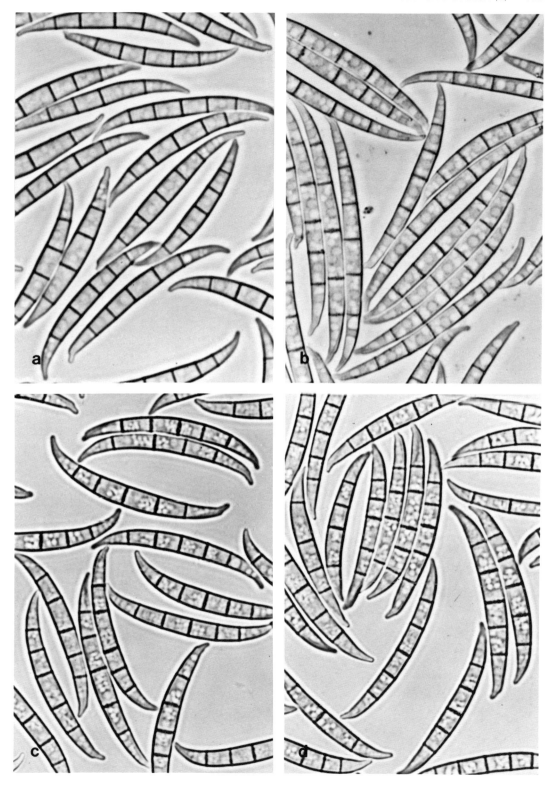

Figure 42. *Fusarium graminearum:* a–d, macroconidia produced in sporodochia (X1000).

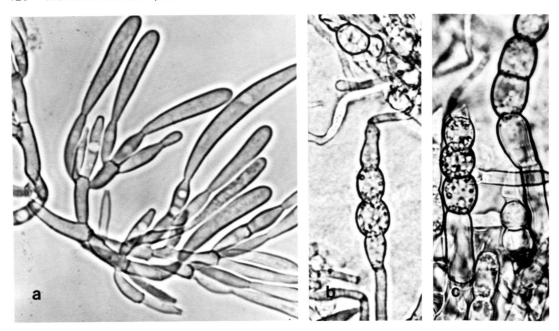

Figure 43. *Fusarium graminearum:* a, conidiophores (monophialides); b, c, chlamydospores (X1000).

8. SECTION DISCOLOR
23. *Fusarium crookwellense* Burgess, Nelson & Toussoun
(This is a new species not included in any taxonomic system; see Ref. 7)

Conidia (Fig. 44). Microconidia are absent. Macroconidia are strongly septate, thick-walled, and unequally curved, with the ventral surface less curved than the dorsal surface, which is strongly arched. The basal cell is distinctly foot-shaped. The apical cell is distinctly curved and tapers to a narrow tip.
Conidiophores (Fig. 45a,b). Unbranched and branched monophialides.
Chlamydospores (Fig. 45c–e). They are present and are formed in the hyphae and the macroconidia.
Perfect state. None known.
Colony morphology. On PDA growth is rapid, with dense aerial mycelium, white in color and then tan. Orange to red-brown sporodochia generally appear early in the center of the culture and later in other portions of the culture. The undersurface is carmine red.

Most distinguishing characteristics. The cultural resemblance to *F. culmorum* on PDA, but with macroconidia that may be confused with *F. graminearum*. However, the macroconidia are shorter and the dorsal surface is more arched than in macroconidia of *F. graminearum* produced on CLA.
Affinities. *Fusarium crookwellense* has the typical rapid, profuse growth and carmine red undersurface on PDA which is characteristic of species in the section Discolor. Culturally on PDA it resembles *F. culmorum*. In the shape of its macroconidia produced on CLA it resembles *F. graminearum*.

Fusarium crookwellense is a rather stable fungus in culture, but mutants do occur, especially pionnotal mutants in which the aerial mycelium is absent and the colony surface is a sheet of macroconidia in pionnotes. Both surfaces are then carmine red.

To date, *Fusarium crookwellense* has been isolated from a variety of substrates in Australia, U.S.A., Republic of South Africa, France, Columbia, and China

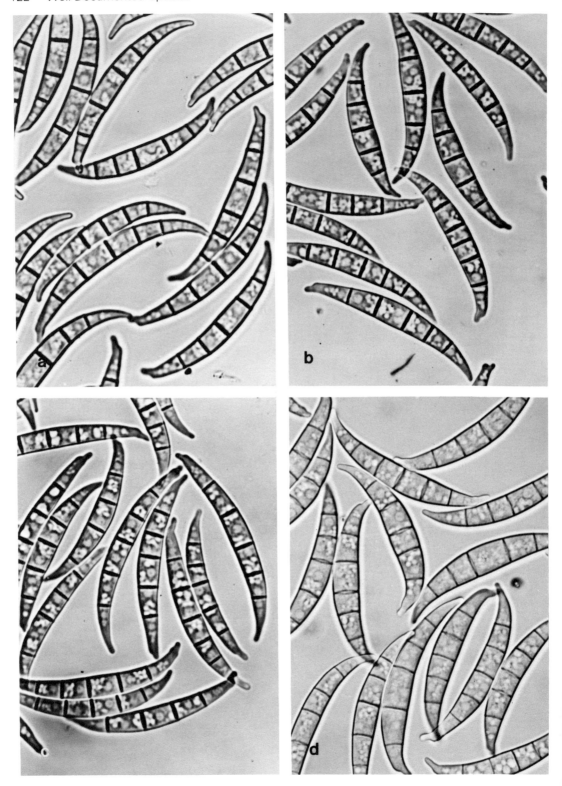

Figure 44. *Fusarium crookwellense:* a–d, macroconidia produced in sporodochia (X1000).

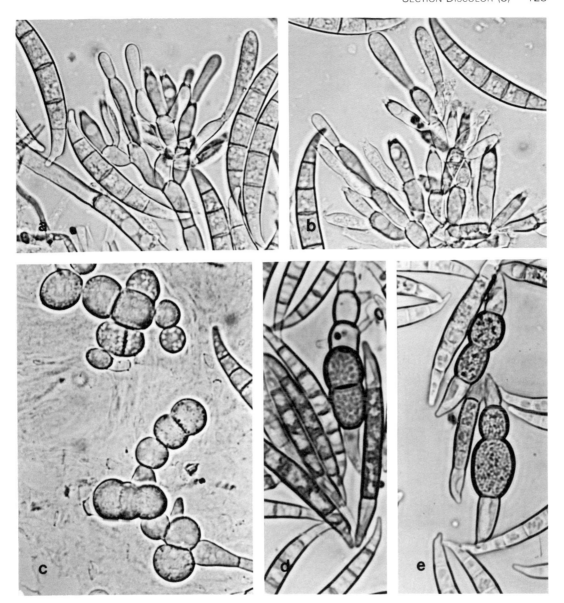

Figure 45. *Fusarium crookwellense:* a, b, conidiophores (monophialides); c–e, chlamydospores. (X1000).

9. SECTION LATERITIUM
24. *Fusarium lateritium* Nees [W&R,G,B,J]
Syn.: *F. stilboides* Wollenw. [W&R,G,B,J]
F. lateritium Nees emend. Snyd. & Hans. [S&H,M&C]

Conidia (Figs. 46, 47, 48a). Microconidia are absent or sparse in most isolates but abundant in some. They are ellipsoidal, spindle- or club-shaped, and 0–3 septate. Macroconidia are abundant and generally cylindrical, with the ventral and dorsal surfaces parallel for most of their length; in some clones the macroconidia are more curved. The apical cell has a characteristic beak or hook. The basal cell is foot-shaped. In some clones, spindle-shaped macroconidia of the *F. semitectum* type may be present (Fig. 47c–e).
Conidiophores (Fig. 48b–d). Unbranched and branched monophialides.
Chlamydospores (Fig. 48e). They are present and are formed singly or in chains. They may be rare in some clones.
Perfect state. *Gibberella baccata* (Wallr.) Sacc. (5).
Colony morphology. On PDA, *F. lateritium* is slow growing, with sparse aerial mycelium. Sporodochia are generally abundant, and usually orange in color. The cultures frequently appear carmine red on both surfaces, but may also be white to pink in color on the upper surface and colorless to light orange to tan on the lower surface.

Most distinguishing characteristics. The shape of the macroconidia with special emphasis on the apical cell and the relative slow growth in culture. Microconidia are absent or sparse in most isolates but abundant in some.
Affinities. None.

The species *F. stilboides* is often separated from *F. lateritium* on the basis of the carmine red color of the cultures. However, in examining a number of cultures designated as *F. stilboides*, we found both red and orange cultures and were unable to separate these cultures from those of *F. lateritium* on the basis of the morphology of the macroconidia.

Wollenweber and Reinking (76) distinguish the varieties *F. lateritium* var. *mori*, var. *minus*, var. *uncinatum*, var. *majus*, and var. *longum*, based on variations in size of macroconidia and colony characteristics. We have grouped these varieties under *F. lateritium*.

Fusarium lateritium is generally stable in culture but may mutate to mycelial forms, which show an increase in aerial mycelium, few sporodochia, and a decrease in colony color. Pionnotal forms also occur as mutants; these cultures have a wet appearance and no aerial mycelium with an orange color on the upper surface. The carmine red undersurface is often partially or completely replaced by the orange of the upper surface.

Fusarium lateritium is frequently found on woody plants.

Fusarium lateritium has been reported to be toxigenic.

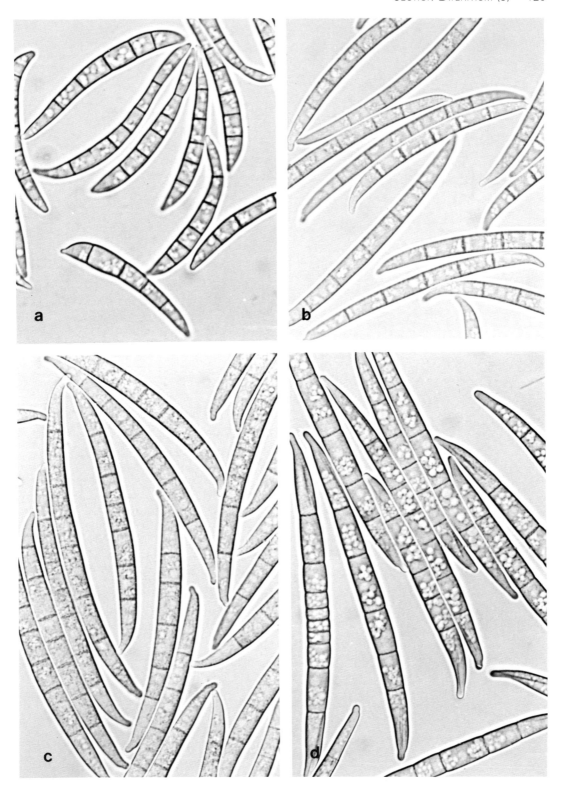

Figure 46. *Fusarium lateritium:* a–d, macroconidia produced in sporodochia (X1000).

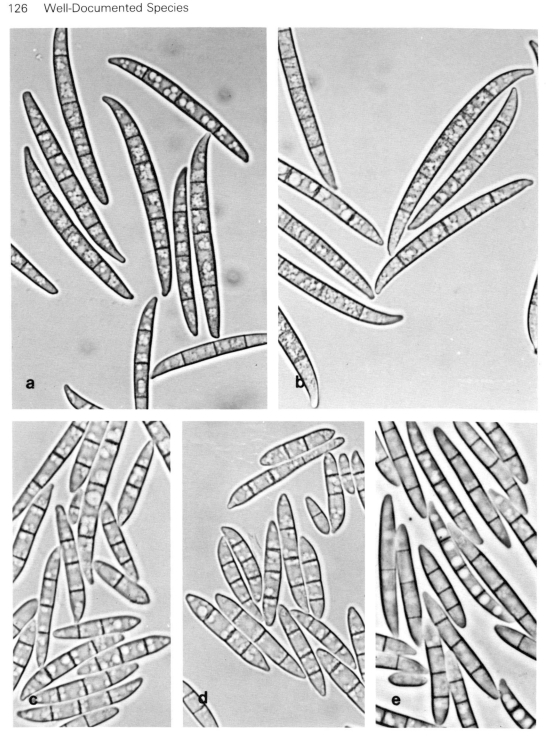

Figure 47. *Fusarium lateritium:* a, b, macroconidia produced in sporodochia; c–e, microconidia produced in the aerial mycelium (X1000).

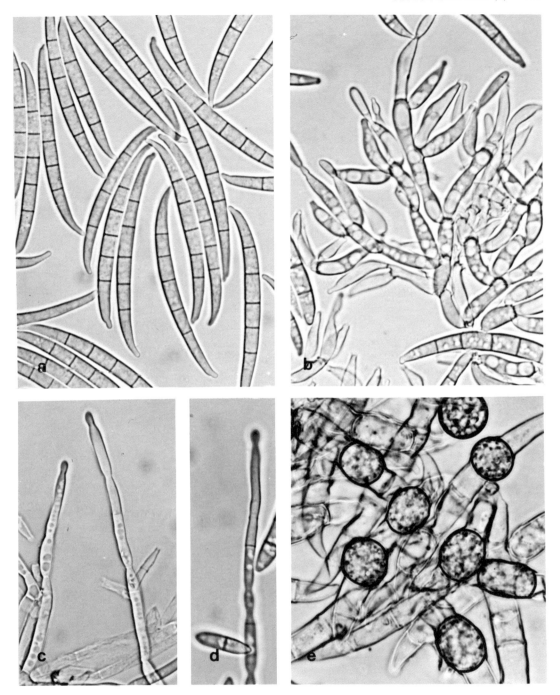

Figure 48. *Fusarium lateritium:* a, macroconidia produced in sporodochia; b–d, conidiophores (monophialides); e, chlamydospores (X1000).

10. SECTION LISEOLA
25. *Fusarium moniliforme* Sheldon [W&R,B,J]

> Syn.: *F. verticillioides* (Sacc.) Nirenberg [G]
> *F. fujikuroi* Nirenberg [G]
> *F. moniliforme* Sheldon emend. Snyd. & Hans. pro parte [S&H,M&C]

Conidia (Figs. 49a–c, 50a,b). Microconidia are abundant and are primarily single-celled, oval to club-shaped, and have a flattened base. They are formed in long chains (Fig. 50d) and in false heads. Macroconidia are present, though sometimes rare. Their appearance varies from slightly sickle-shaped to almost straight with the dorsal and ventral surfaces almost parallel, and they have thin delicate walls. The basal cell is foot-shaped.
Conidiophores (Figs. 49d, 50c,e). Unbranched and branched monophialides.
Chlamydospores. Absent.
Perfect state. *Gibberella fujikuroi* (Sawada) Wollenw. (76).
Colony morphology. On PDA, *F. moniliforme* resembles *F. oxysporum*. Its white aerial mycelium grows rapidly and often becomes tinged with purple. Sporodochia may be present or absent; when present, they may be tan to orange discrete sporodochia, or they may be confluent as pseudo-pionnotes. Sclerotia may also develop and are usually dark blue; they may be so abundant as to give this coloration to the upper and lower surfaces of the colony. The undersurface varies from colorless to dark purple.

Most distinguishing characteristics. Microconidia formed in chains on monophialides and the absence of chlamydospores. Microconidia formed on monophialides in false heads are also present.
Affinities. Members of Wollenweber and Reinking's section Liseola (76), which includes *F. moniliforme,* resemble *F. oxysporum* on PDA and frequently can be confused with this species.

It is important to note that the length and number of chains can vary greatly from culture to culture. In some cases they may be rare and the chains may consist of as few as three conidia; nevertheless these are chains and not false heads.

Refer also to the discussion of the use of the KCl medium (page 14).

We have retained the name *F. moniliforme* Sheldon as used by Wollenweber and Reinking (76), but have included *F. moniliforme* Sheldon var. *minus* Wollenw. which was based on the absence of sporodochia and pionnotes. We have not accepted the combination *F. verticillioides* (Sacc.) Nirenberg, although this may well be an older valid name (45), because we think that the International Code of Botanical Nomenclature with respect to priority is difficult or impossible to apply to most species of *Fusarium* in the absence of type specimens. We agree with Booth's statement (4) that "it would be preferable to conserve all the names now in use rather than adopt a wholesale series of name changes in an effort to bring nomenclature strictly within the type system." Consequently we have retained the well-established name *F. moniliforme*. We have also not accepted *F. fujikuroi* Nirenberg as a separate species and have included this in *F. moniliforme* on the basis of our observations on the variability in chain length in different isolates of *F. moniliforme* and because we could not find true polyphialides as described by Nirenberg (47) in authentic cultures of *F. fujikuroi*. These cultures are characterized by short microconidial chains, but on KCl medium they produce much longer chains on monophialides and are indistinguishable from *F. moniliforme*. We refer to them as "short-chained" types of *F. moniliforme*.

A taxonomic treatment different from that of Wollenweber and Reinking (76) has been proposed by Nirenberg (47).

Fusarium moniliforme varies in culture to pionnotal forms where the aerial mycelium is suppressed and replaced by sheets of macroconidia that give the culture a yellowish, wet appearance, and to mycelial forms where sporodochia and colors of the colony are suppressed or eliminated.

Fusarium moniliforme is cosmopolitan.

Fusarium moniliforme has been reported to be toxigenic.

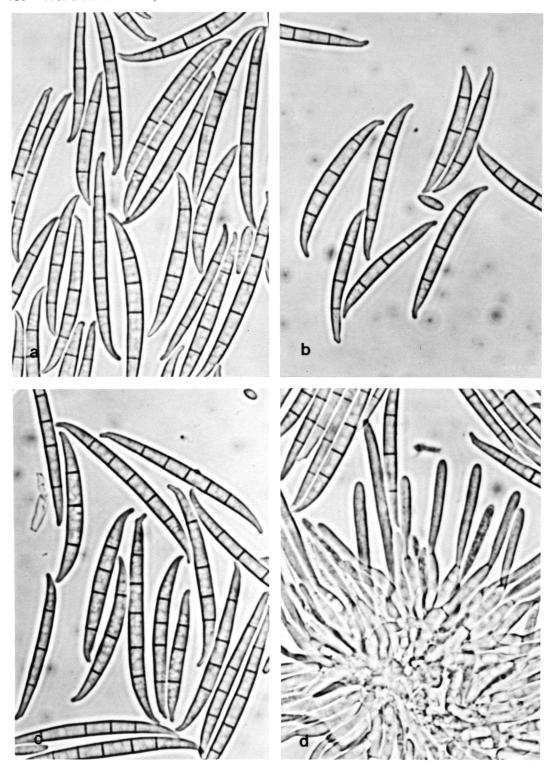

Figure 49. *Fusarium moniliforme:* a–c, macroconidia produced in sporodochia; d, macroconidio-phores (monophialides) (X1000).

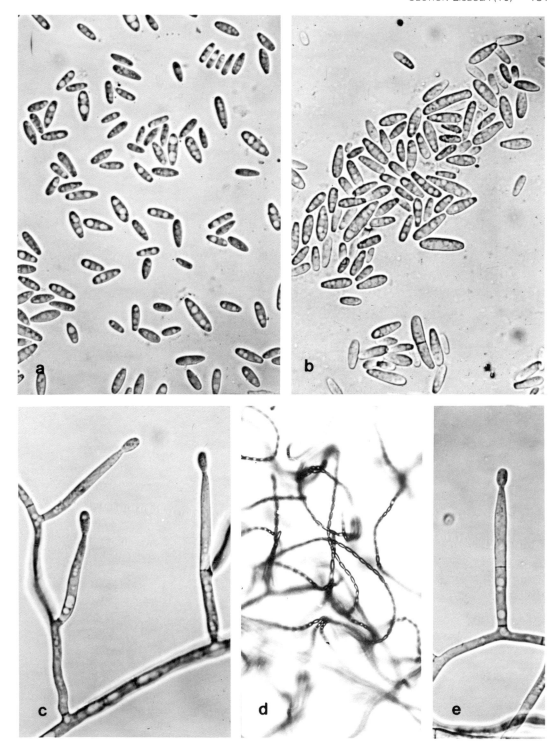

Figure 50. *Fusarium moniliforme:* a, b, microconidia produced in the aerial mycelium; c, e, micro-conidiophores (monophialides); d, microconidia in chains (a–c, e X1000; d X190).

10. SECTION LISEOLA

26. *Fusarium proliferatum* (Matsushima) Nirenberg (47) [G]

Syn.: *F. moniliforme* Sheldon pro parte [W&R,B,J]

F. moniliforme Sheldon emend. Snyd. & Hans. pro parte [S&H,M&C]

Conidia (Figs. 51a,b,d, 52a,b). Microconidia are abundant, usually single-celled or club-shaped with a flattened base. Pear-shaped microconidia may also occur but generally are rare. Microconidia are borne in chains of varying length (Fig. 52c) and in false heads. Macroconidia are abundant, only slightly sickle-shaped to almost straight, with the dorsal and ventral surfaces parallel for most of the length of the macroconidium. The walls are thin and delicate and the basal cell is foot-shaped.

Conidiophores (Figs. 51c, 52d,e). Unbranched and branched polyphialides and monophialides.

Chlamydospores. Absent.

Perfect state. None known.

Colony morphology. On PDA, *F. proliferatum* resembles *F. oxysporum*. The white aerial mycelium grows rapidly and is sometimes tinged with purple. Sporodochia may be present or absent; when present, they may be tan to orange discrete sporodochia, or they may be confluent as pseudo-pionnotes. Sclerotia may also develop and are frequently dark-blue; they may be so abundant that they give this coloration to the colony surface and undersurface. The undersurface varies from colorless to dark purple.

Most distinguishable characteristics. Microconidia are formed in chains on polyphialides, and the presence of polyphialides separates *F. proliferatum* from *F. moniliforme*. The chains of *F. proliferatum* on the polyphialides often appear in the shape of a 'V' under the low-power objective of the microscope. Chlamydospores are absent.

Affinities. Members of Wollenweber and Reinking's section Liseola (76), which includes *F. proliferatum,* resemble *F. oxysporum* on PDA and can be frequently confused with this species.

It is important to note that the length and the number of chains of microconidia can vary greatly from culture to culture. In some cases they may be rare and consist of as few as three conidia; nevertheless these are chains, not false heads.

Refer also to the discussion of the use of the KCl medium (page 14).

A taxonomic treatment different from that of Wollenweber and Reinking for the section Liseola (76) has been proposed by Nirenberg (47).

Fusarium proliferatum varies in culture to pionnotal forms where the aerial mycelium is suppressed and replaced by sheets of macroconidia that give the culture a yellowish, wet appearance, and to mycelial forms where sporodochia and colors of the colony are suppressed or eliminated.

Fusarium proliferatum is cosmopolitan.

Fusarium proliferatum has been reported to be toxigenic.

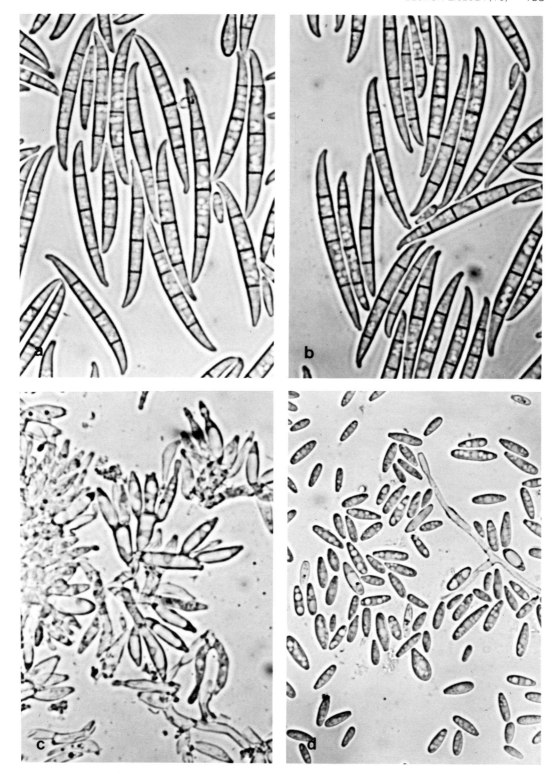

Figure 51. *Fusarium proliferatum:* a, b, macroconidia produced in sporodochia; c, conidiophores (monophialides); d, microconidia produced in the aerial mycelium (X1000).

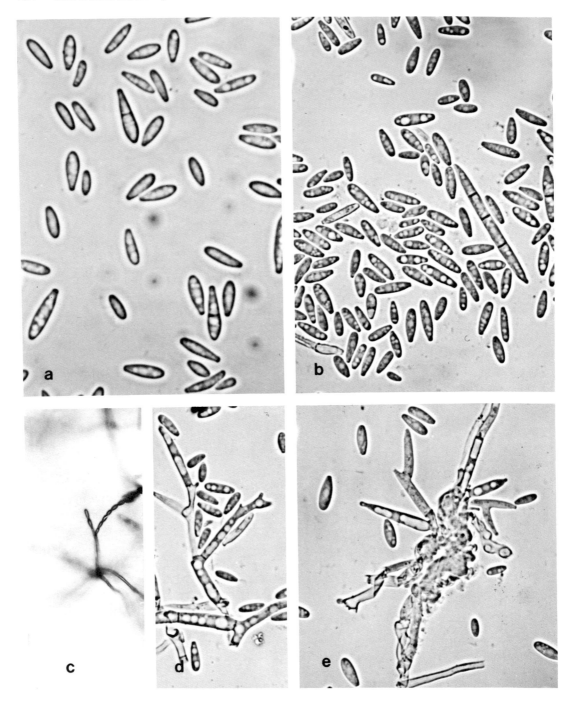

Figure 52. *Fusarium proliferatum:* a, b, microconidia produced in the aerial mycelium; c, microconidia in chains formed from a polyphialide; d, e, polyphialides (a, b, d, e X1000; c X190).

10. SECTION LISEOLA

27. *Fusarium subglutinans* (Wollenw. & Reinking) Nelson, Toussoun & Marasas comb. nov.

Basionym: *Fusarium moniliforme* Sheldon var. *subglutinans* Wollenw. & Reinking, Phytopathology 15:163. 1925. [W&R,B,J]

> Syn.: *F. sacchari* (Butler) Gams var. *subglutinans* (Wollenw. & Reinking) Niren-
> berg [G]
>
> *F. moniliforme* Sheldon emend. Snyd. & Hans. 'Subglutinans' sensu Snyd.,
> Hans. & Oswald (64) [S&H,M&C]

The main differences between *F. subglutinans* and *F. moniliforme* that require elevation of *F. moniliforme* var. *subglutinans* to species rank are the absence of microconidial chains and the presence of polyphialides in *F. subglutinans*. This species differs from the closely related *F. anthophilum* (A. Braun) Wollenw. in that pyriform microconidia are absent.

Conidia (Figs. 53a–c, 54a,b). Microconidia are abundant, oval, and usually single-celled, but may be 1–3 septate. Microconidia are produced only in false heads. Macroconidia are abundant, only slightly sickle-shaped to almost straight with the dorsal and ventral surfaces almost parallel, and with thin, delicate walls. The basal cell is foot-shaped.

Conidiophores (Figs. 53d, 54c–e). Unbranched and branched polyphialides and mono-phialides.

Chlamydospores. Absent.

Perfect state. *Gibberella subglutinans* (Edwards) Nelson, Toussoun & Marasas comb. nov. Basionym: *Gibberella fujikuroi* (Sawada) Wollenw. var. *subglutinans* Edwards, Agric. Gaz. New South Wales 44:896. 1933. [W&R,B,G]

The ascospores are mostly 1 septate but may become 2 to 4 septate and are somewhat narrower (14–18 × 4.5–5.0 μm) than those of *G. fujikuroi* (Sawada) Wollenw. [compare Booth (4) and Wollenweber and Reinking (76)]. Both species are heterothallic (4), but we have recently confirmed the observation by Ullstrup (70) that certain strains of *G. subglutinans* are homothallic. A paper dealing with these homothallic strains of *G. subglutinans* is in preparation.

Colony morphology. On PDA, *F. subglutinans* resembles *F. oxysporum*. The white aerial mycelium grows rapidly and is sometimes tinged with purple. Sporodochia may be present or absent; when present, they may be tan to orange discrete sporodochia, or they may be confluent as pseudo-pionnotes. Sclerotia may also develop and are frequently dark blue. They may be so abundant as to give this coloration to the colony surface and undersurface. The undersurface varies from colorless to dark purple.

Most distinguishing characteristics. Microconidia formed on polyphialides, but always in false heads and never in chains. Chlamydospores are absent.

Affinities. Members of Wollenweber and Reinking's section Liseola (76), which includes *F. subglutinans*, resemble *F. oxysporum* on PDA and frequently can be confused with this species.

Refer also to the discussion of the use of the KCl medium (page 14).

A taxonomic treatment different from that of Wollenweber and Reinking (76) has been proposed by Nirenberg (47).

Fusarium subglutinans varies in culture to pionnotal forms in which the aerial mycelium is suppressed and replaced by sheets of macroconidia that give the culture a yellowish, wet appearance, and to mycelial forms in which the sporodochia and colony colors are suppressed or eliminated.

Fusarium subglutinans is cosmopolitan.

Fusarium subglutinans has been reported to be toxigenic.

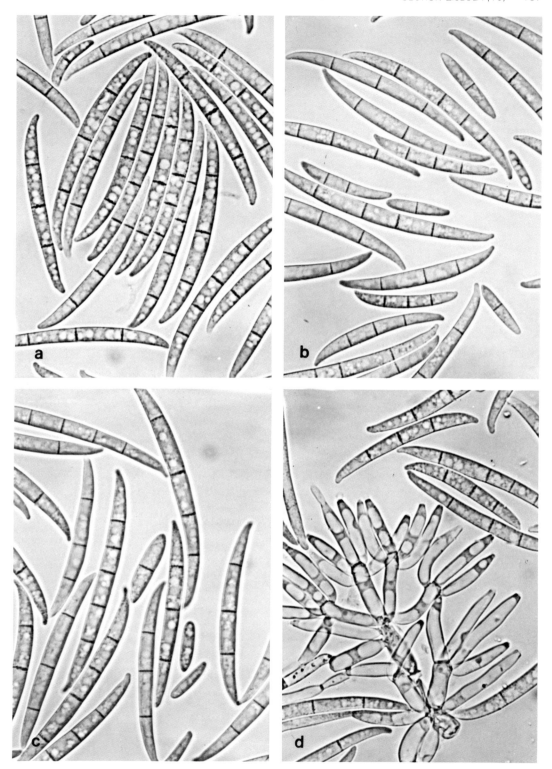

Figure 53. *Fusarium subglutinans:* a–c, macroconidia produced in sporodochia; d, macroconidio-phores (monophialides) (X1000)

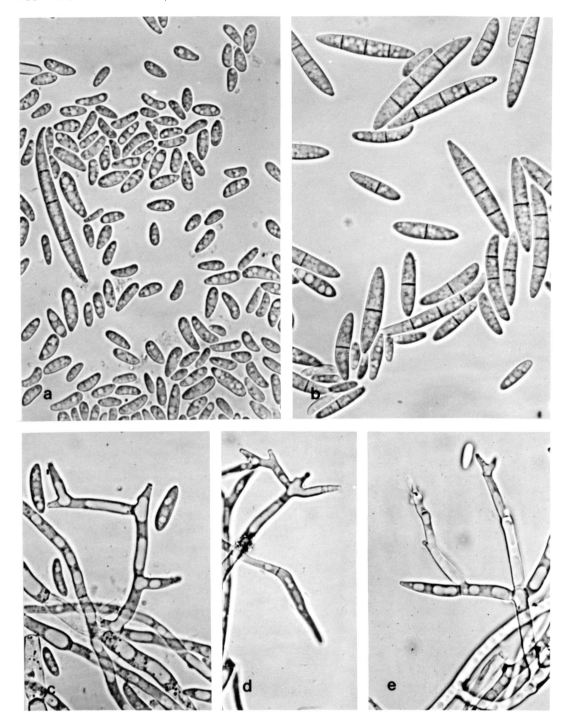

Figure 54. *Fusarium subglutinans:* a, b, microconidia produced in aerial mycelium; c–e, polyphialides (X1000).

10. SECTION LISEOLA
28. *Fusarium anthophilum* (A. Braun) Wollenw. (47) [G]
Syn.: *F. moniliforme* Sheldon var. *anthophilum* (A. Braun) Wollenw. [W&R,J]
F. moniliforme Sheldon pro parte [B]
F. moniliforme Sheldon emend. Snyd. & Hans. pro parte [S&H, M&C]

Conidia (Figs. 55, 56a,b,e). Microconidia are abundant and may be single-celled, oval, globose, pear-shaped, or club-shaped. They are produced only in false heads. Macroconidia are abundant, only slightly sickle-shaped to almost straight with the dorsal and ventral surfaces almost parallel, and have thin, delicate walls. The basal cell is foot-shaped.
Conidiophores (Fig. 56c–f). Unbranched and branched polyphialides and monophialides.
Chlamydospores. Absent.
Perfect state. None known.
Colony morphology. On PDA, *F. anthophilum* resembles *F. oxysporum.* The aerial mycelium is white in color and grows rapidly. Sporodochia may be present or absent; when present, they may be tan to orange discrete sporodochia, or they may be confluent as pseudo-pionnotes. Sclerotia may be present. The undersurface of the culture varies from colorless to purple.

Most distinguishing characteristics. Oval, globose, or pear-shaped microconidia borne on polyphialides. These distinctive microconidia separate *F. anthophilum* from *F. subglutinans.* Microconidia are not formed in chains. Chlamydospores are absent.
Affinities. Members of Wollenweber and Reinking's section Liseola (76), which includes *F. anthophilum,* resemble *F. oxysporum* on PDA and frequently can be confused with this species.

Refer also to the discussion of the use of the KCl medium (page 14).

A taxonomic treatment different from that of Wollenweber and Reinking's section Liseola (76) has been proposed by Nirenberg (47).

Fusarium anthophilum varies in culture to pionnotal forms where the aerial mycelium is suppressed and replaced by sheets of macroconidia that give the cultures a yellowish, wet appearance, and to mycelial forms where sporodochia and colony colors are suppressed or eliminated.

We have encountered *F. anthophilum* frequently in the U.S.A and in the tropics.

Fusarium anthophilum has been reported to be toxigenic.

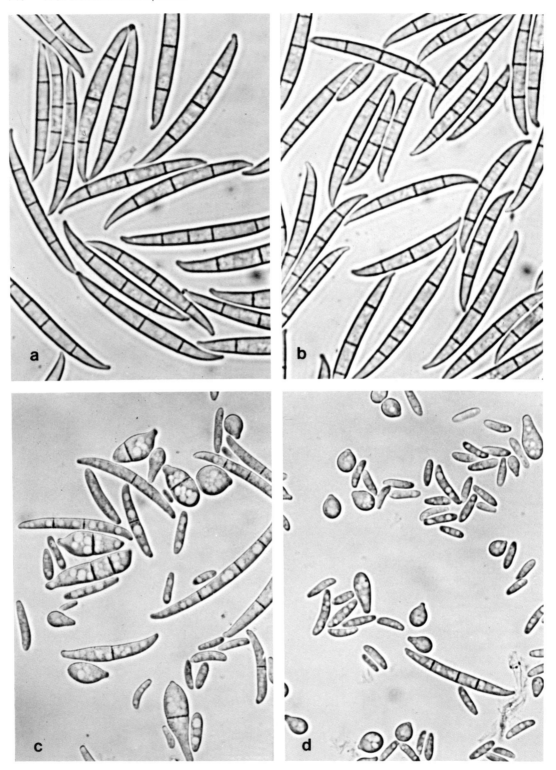

Figure 55. *Fusarium anthophilum:* a, b, macroconidia produced in sporodochia; c, d, macroconidia and microconidia produced in the aerial mycelium (X1000).

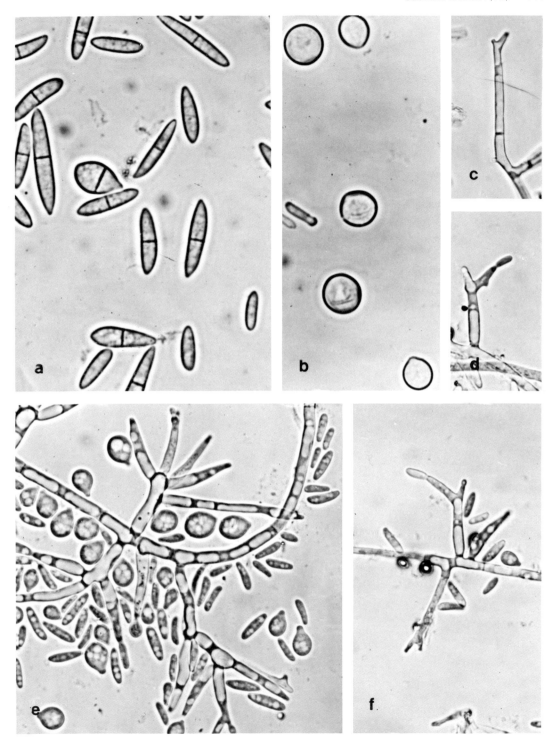

Figure 56. *Fusarium anthophilum:* a, b, microconidia produced in the aerial mycelium; c–f, poly-phialides and monophialides (X1000).

11. SECTION ELEGANS
29. *Fusarium oxysporum* Schlecht. emend. Snyd. & Hans. [S&H,M&C]

Syn.: All species in section Elegans [W&R]

F. oxysporum Schlecht. emend. Snyd. & Hans. pro parte [G,B]

F. redolens Wollenw. [G]

F. oxysporum Schlecht. emend. Snyd. & Hans. var. *redolens* (Wollenw.) Gordon [B]

F. oxysporum Schlecht. [J]

F. oxysporum Schlecht. var. *redolens* (Wollenw.) Gordon [J]

Conidia (Figs. 57, 58a,b, 59c,d). Microconidia are abundant, generally single-celled, oval to kidney-shaped, and produced only in false heads. Macroconidia are abundant, only slightly sickle-shaped, thin-walled, and delicate, with an attenuated apical cell and a foot-shaped basal cell.

Conidiophores (Fig. 59a,b,e–g). Unbranched and branched monophialides. The monophialides bearing microconidia are short when compared to those produced by *F. moniliforme* and *F. solani.*

Chlamydospores (Fig. 58c–e). They are present and are formed singly or in pairs. In most isolates they form readily and profusely in culture.

Perfect state. None known.

Colony morphology. On PDA, growth is rapid and the white aerial mycelium may become tinged with purple or be submerged by the blue color of the sclerotia when they are abundant, especially at the base of the slant, or by the cream to tan to orange sporodochia when these are abundant. Discrete, erumpent orange sporodochia are present in some strains. The undersurface may be colorless to dark blue or dark purple, and these colors may be visible through the mycelium when viewed from above.

Most distinguishing characteristics. The presence of chlamydospores, and microconidia borne in false heads on short monophialides. Cultures of *F. oxysporum* often resemble cultures of *F. subglutinans,* but the latter has microconidia borne on polyphialides and chlamydospores are absent.

Affinities. On PDA *F. oxysporum* resembles the species in Wollenweber and Reinking's section Liseola (76).

Wollenweber and Reinking (76) and Gerlach (21) recognize *F. redolens* as a species, while Booth (4) recognizes *F. oxysporum* var. *redolens.* We have placed both in *F. oxysporum.*

Fusarium oxysporum has many clones which exhibit a variety of morphological features on PDA. In addition it mutates frequently in culture, either to forms that progressively become more mycelial with an increase in aerial mycelium and a decrease in sporodochia, sclerotia, and color, or to pionnotal forms in which the aerial mycelium is depressed, and macroconidia are borne in pionnotes that give the cultures a wet, yellow to orange appearance (72).

Fusarium oxysporum is cosmopolitan.

Fusarium oxysporum has been reported to be toxigenic.

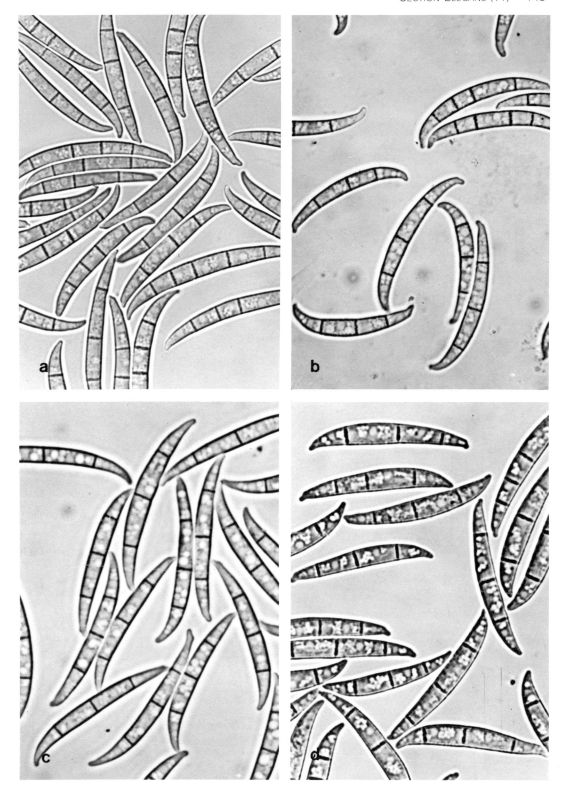

Figure 57. *Fusarium oxysporum:* a–d, macroconidia produced in sporodochia (X1000).

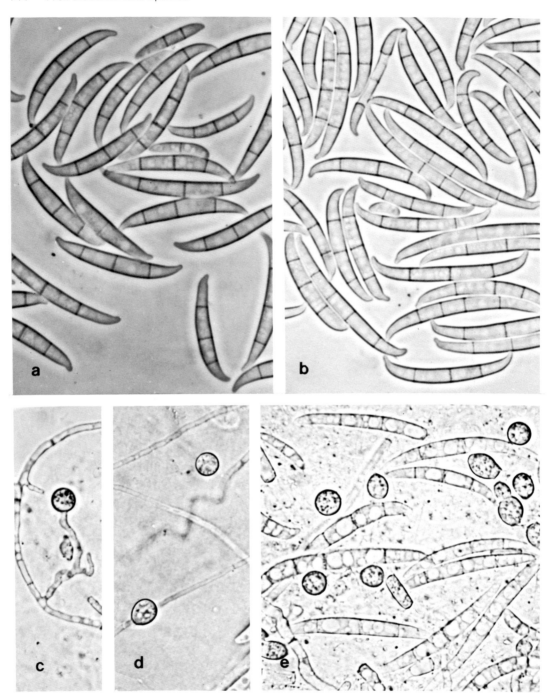

Figure 58. *Fusarium oxysporum:* a–b, macroconidia produced in sporodochia; c–e, chlamydospores (X1000).

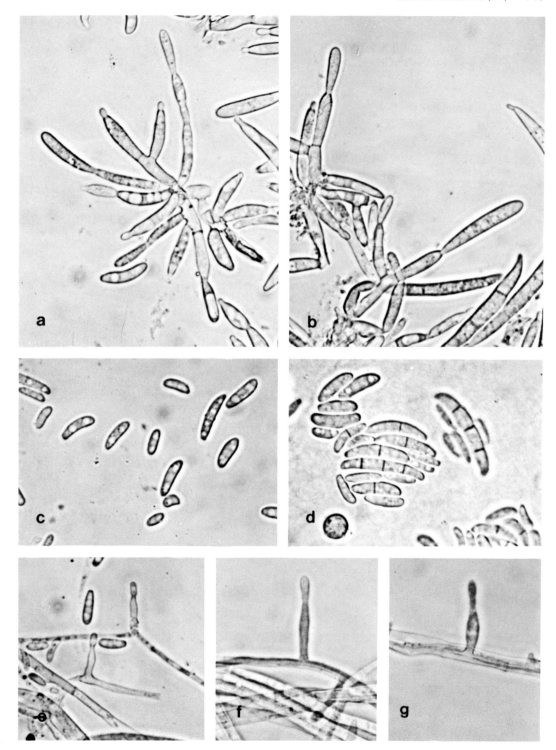

Figure 59. *Fusarium oxysporum:* a, b, macroconidiophores (monophialides); c, d, microconidia produced in the aerial mycelium; e–g, microconidiophores (monophialides) (X1000).

12. SECTIONS MARTIELLA and VENTRICOSUM

30. *Fusarium solani* (Mart.) Appel & Wollenw. emend. Snyd. & Hans. [S&H,M&C]

Syn.: All species in sections Martiella and Ventricosum [W&R]

F. javanicum Koorders [G,J]

F. coeruleum (Libert) Sacc. [G]

F. solani (Mart.) Sacc. [G,B,J]

F. eumartii Carpenter [G]

F. illudens Booth [G,B]

F. ventricosum Appel & Wollenw. [G,B]

F. solani (Mart.) Sacc. var. *coeruleum* (Sacc.) Booth [B]

F. tumidum Sherb. [B]

F. solani (Mart.) Sacc. var. *coeruleum* (Libert) Bilai [J]

F. solani (Mart). Sacc. var. *ventricosum* (Appel & Wollenw.) Joffe [J]

Conidia (Figs. 60, 61, 62c–e). Microconidia are present, varying from sparse to abundant, generally single-celled, oval to kidney-shaped. The microconidia are similar in shape to those found in *F. oxysporum,* but they are larger and have thicker walls. Macroconidia are abundant, stout, thick-walled, and generally cylindrical, with the dorsal and ventral surfaces parallel for most of their length. The apical cell is blunt and rounded, and the basal cell is rounded or is distinctly foot-shaped or notched.

Conidiophores (Figs. 62a,b, 63a,b). Unbranched and branched monophialides. The monophialides bearing microconidia are long when compared to those in *F. oxysporum.*

Chlamydospores (Fig. 63c–e). They are present and are formed singly and in pairs. They are abundant in most clones.

Perfect state. *Nectria haematococca* Berk. & Br. (5).

Colony morphology. On PDA growth is rapid, often with abundant aerial mycelium. The surface is soon covered with confluent sporodochia that give the appearance of pionnotes and color the surface cream, blue-green, or blue, but never orange. Some clones may show a dark purple color on the upper surface. The undersurface is generally colorless, but some clones produce a dark violet pigment.

Distinguishing characteristics. The morphology of the macroconidia, the elongate monophialides bearing microconidia, which also help distinguish it from *F. oxysporum,* and the distinctive cream, blue-green or blue color of colonies on PDA.

Affinities. None.

Gerlach (21) and Booth (4) recognize several species and varieties which we group in *F. solani.*

Fusarium solani is stable in culture but can mutate to mycelial types with abundant aerial mycelium, no sporodochia, and no color.

Fusarium solani is cosmopolitian.

Fusarium solani has been reported to be toxigenic.

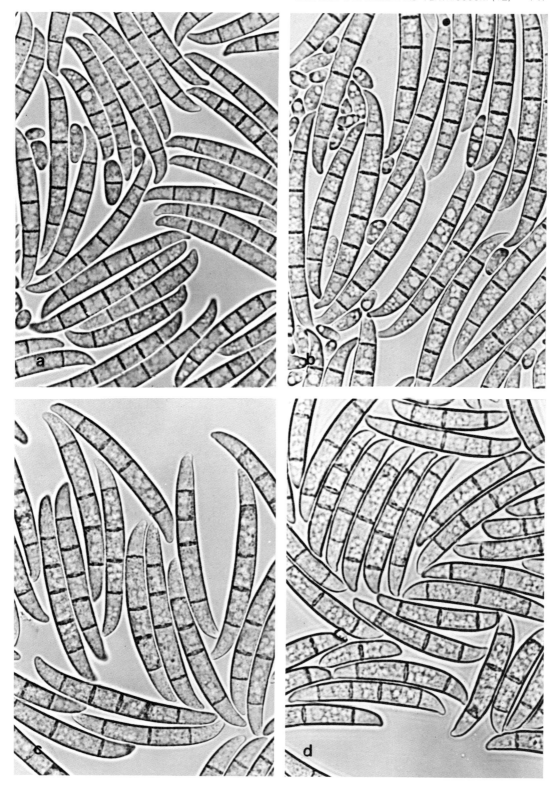

Figure 60. *Fusarium solani:* a–d, macroconidia produced in sporodochia (X1000).

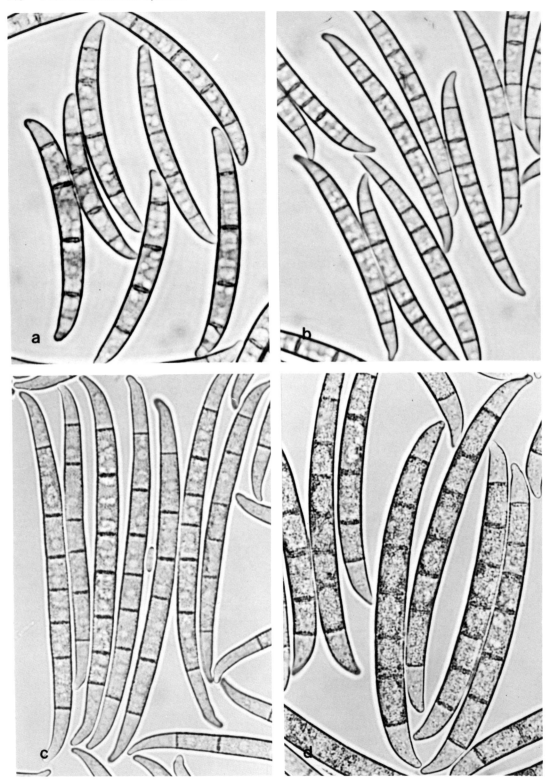

Figure 61. *Fusarium solani:* a–d, macroconidia produced in sporodochia (X1000).

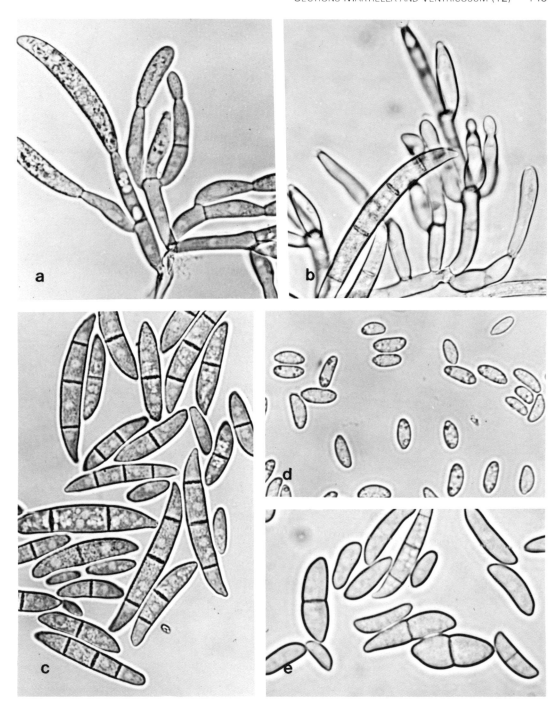

Figure 62. *Fusarium solani:* a, b, macroconidiophores (monophialides); c–e, microconidia produced in the aerial mycelium (X1000).

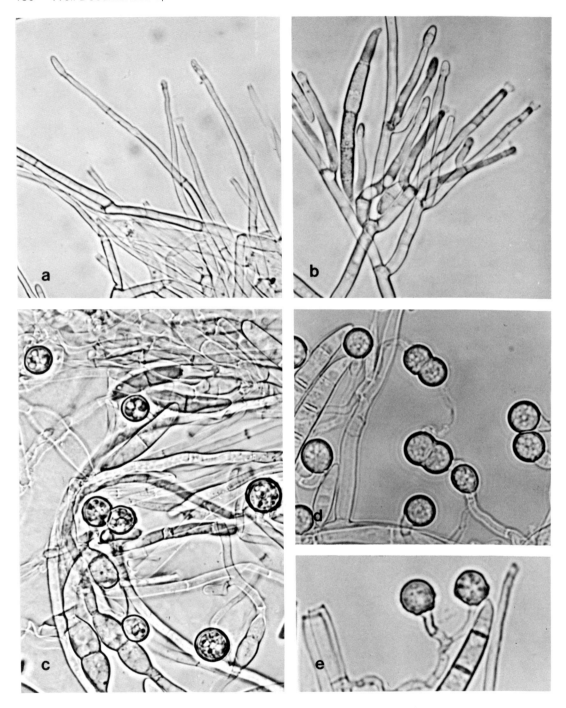

Figure 63. *Fusarium solani:* a, b, microconidiophores (monophialides); c–e, chlamydospores (X1000).

Descriptions and Illustrations of Insufficiently Documented *Fusarium* Species

Fusarium species which we have not seen, or of which we have only two or three cultures sent to us by other workers, are listed below. The species we have cultures of are marked with an asterisk, and brief descriptions and illustrations of these species are presented.

Section Eupionnotes
   F. cavispermum
   F. melanochlorum
   F. tabacinum
   F. epistroma
Section Macroconia
   F. buxicola
   F. sphaeriae
   *F. coccophilum**
   F. gigas
Section Submicrocera
   *F. ciliatum**
Section Pseudomicrocera
   *F. coccidicola**

Section Arachnites
  F. stoveri
  F. larvarum *
Section Roseum
  F. detonianum *
Section Arthrosporiella
  F. diversisporum *
Section Discolor
  F. flocciferum *
  F. sublunatum *
  F. tumidum *
  F. robustum *
  F. buharicum *
  F. lunulosporum *
Section Lateritium
  F. xylarioides *
Section Liseola
  F. annulatum *
  F. succisae *
Sections Elegans or Lateritium
  F. udum *

The species presented in the following pages are treated briefly because we have seen a very limited number of specimens (generally 1–3 isolates). Consequently we cannot present a complete treatment of these species. We have limited ourselves to a summary description taken from the literature together with photographs of the conidia and conidiophores from the cultures we have examined. Many of these cultures have been kept in collections for extended periods and have degenerated to varying degrees through repeated subculturing; they therefore may not be representative of specimens existing in nature. Because of the unfavorable nature of the material both in quantity and quality, the species presented in the following pages have not been included in the synoptic keys.

We repeat here our firm opinion that a taxonomic treatment, to be adequate, must be based on the observations of an extensive number of specimens taken from as wide a geographic distribution as possible in order to understand the boundaries of natural variation. This examination, together with a study of cultural mutants in the laboratory, should form the basis for speciation. Until such time as this is accomplished, the species presented on the following pages remain questionable in our opinion.

We are limiting ourselves to four taxonomic systems in the cross-referencing in this section. The taxonomic systems used are those proposed by Wollenweber and Reinking (76), Gerlach (21), Booth (4), and Snyder and Hansen (44, 58, 59, 60, 65, 68). Many of the *Fusarium* species dealt with in this section are not mentioned by Joffe (35); when Joffe does discuss them, he follows the same system as Wollenweber and Reinking (76) or Gerlach (21). Likewise Messiaen and Cassini (41) follow the system proposed by Snyder and Hansen (60) for these species. For this reason both the taxonomic systems proposed by Joffe (35) and Messiaen and Cassini (41) are omitted in the cross-referencing of this group of *Fusarium* species.

SECTION MACROCONIA
Fusarium coccophilum (Desm.) Wollenw. & Reinking [W&R,G,B]
Syn.: *F. episphaeria* (Tode) Snyd. & Hans. pro parte [S&H]

Perfect state. *Nectria flammea* (Tulasne) Dingley (4, 21).
Cultures are slow growing with prominent orange sporodochia. The sporodochia may be arched and synnema-like. Microconidia are absent. Macroconidia (Fig. 64) are long and cylindrical, with the apical cell narrowed or bent in a hook. The basal cell is foot-shaped. Chlamydospores are absent.

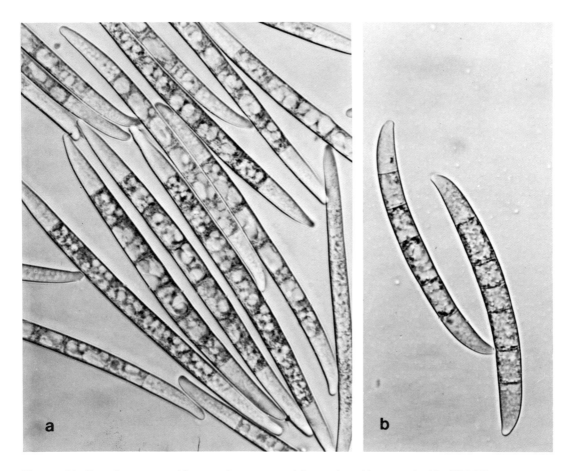

Figure 64. *Fusarium coccophilum:* a, b, macroconidia produced in sporodochia (X1000).

SECTION SUBMICROCERA
Fusarium ciliatum Lk. [W&R,G]
      Syn.: *F. aquaeductum* Lagerh. pro parte [B]
      [Listed as a doubtful species by Snyder and Hansen]

Perfect state. *Calonectria decora* (Wallr.) Sacc. (21).
Cultures are slow growing with prominent orange sporodochia which may be arched and synnema-like. Microconidia are absent. Macroconidia (Fig. 65) are long and very slender with thin walls. The basal cell is not foot-shaped. Chlamydospores are absent.

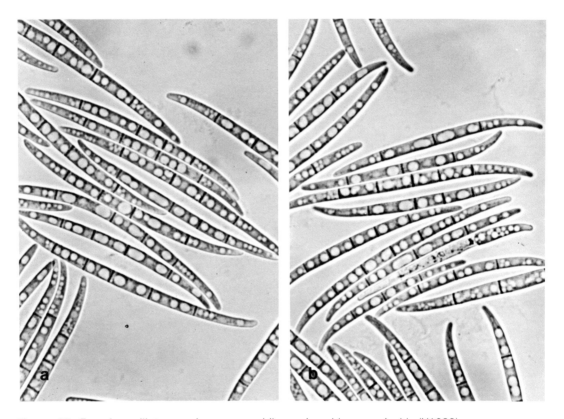

Figure 65. *Fusarium ciliatum:* a, b, macroconidia produced in sporodochia (X1000).

SECTION PSEUDOMICROCERA
Fusarium coccidicola P. Henn. [G]
Syn.: *F. juruanum* P. Henn. [W&R,B]
[Not mentioned by Snyder & Hansen]

Perfect state. *Calonectria diploa* (Berk. & Curt.) Wollenw. (21).
Cultures are slow growing and orange sporodochia develop, sometimes in the shape of synnemata. Microconidia are absent. Macroconidia (Fig. 66a,b) are long, thin, and curved in a sickle-shape with pointed ends. The basal cell may be foot-shaped. Conidiophores (Fig. 66c,d) are unbranched or branched monophialides. Chlamydospores are absent.

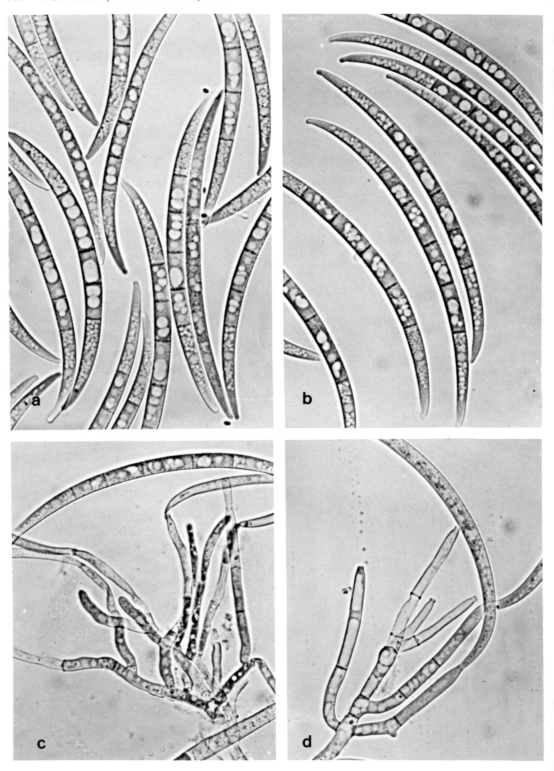

Figure 66. *Fusarium coccidicola:* a, b, macroconidia produced in sporodochia; c, d, conidiophores (monophialides) (X1000).

SECTION ARACHNITES

Fusarium larvarum Fuckel [W&R,G,B]

Syn.: *F. nivale* Fr. emend. Snyd. & Hans. pro parte [S&H]

Perfect state. *Nectria aurantiicola* Berk. & Br. (5).
Cultures are slow growing, with orange sporodochia formed in the white aerial mycelium. The undersurface is white to tan. Microconidia are absent. Macroconidia (Fig. 67) are curved, 1–3 septate, and the basal cell is not foot-shaped. Chlamydospores are absent.

Booth (4) transferred *F. larvarum* from the section Arachnites to the new section Coccophilum together with two other hypocreaceous *Fusarium* species parasitic on scale insects. Although we agree that the affinities of *F. larvarum* are probably with these other entomogenous *Fusarium* species rather than with *F. nivale,* we are not well acquainted with *F. larvarum* or its perfect state and are retaining the classification of Wollenweber and Reinking (76) for the purpose of this volume.

Fusarium larvarum has been reported to be toxigenic.

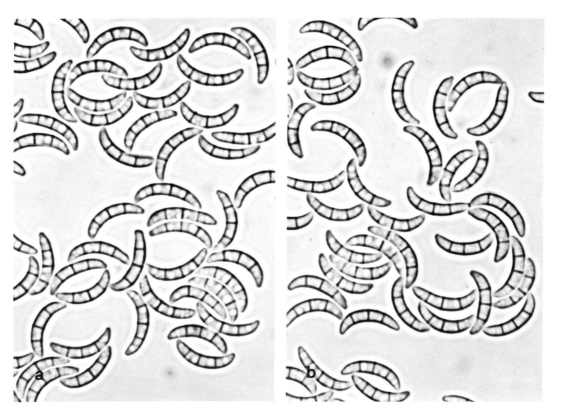

Figure 67. *Fusarium larvarum:* a, b, macroconidia produced in sporodochia (X1000).

SECTION ROSEUM
Fusarium detonianum Sacc. [W&R,G]

Syn.: *F. avenaceum* (Corda ex Fr.) Sacc. pro parte [B]

F. roseum Lk. emend. Snyd. & Hans. 'Avenaceum' pro parte [S&H]

Perfect state. None known.

Wollenweber and Reinking (76) separated this species from *F. avenaceum* because of the longer macroconidia shown in Fig. 68. We include this species in *F. avenaceum*.

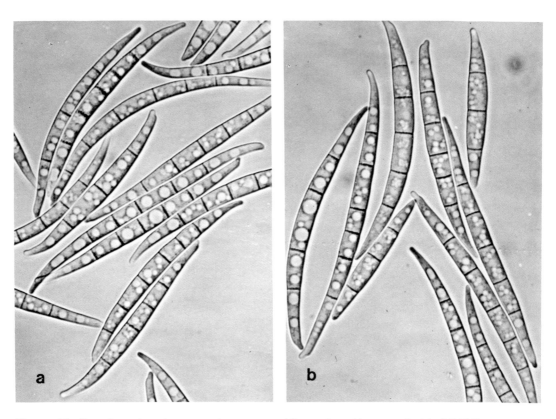

Figure 68. *Fusarium detonianum:* a, b, macroconidia produced in sporodochia (X1000).

SECTION ARTHROSPORIELLA

Fusarium diversisporum Sherb. [W&R,G]

Syn.: *F. semitectum* (Berk. & Rav.) var. *majus* Wollenw. pro parte [B]
F. roseum Lk. emend. Snyd. & Hans. pro parte [S&H]

Perfect state. None known.

Wollenweber and Reinking (76) separated this species from *F. semitectum* on the basis of the diverse shapes of the conidia, which vary from the typical *F. semitectum* shape to sickle-shaped to "S" shaped. The colony appearance is similar to *F. semitectum*. Macroconidia and a conidiophore (polyphialide) are illustrated in Fig. 69. We include this species in *F. semitectum*.

Fusarium diversisporum has been reported to be toxigenic.

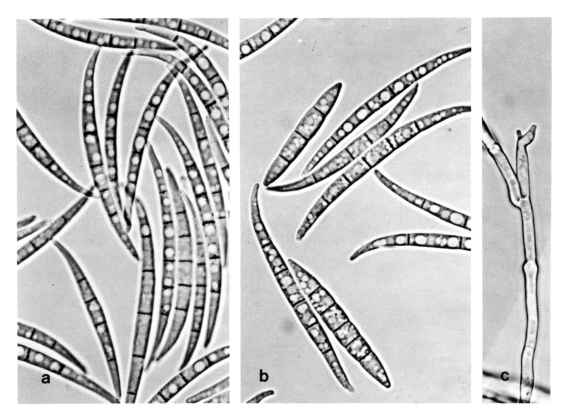

Figure 69. *Fusarium diversisporum:* a, macroconidia produced in sporodochia; b, macroconidia produced in the aerial mycelium; c, polyphialide (X1000).

SECTION DISCOLOR
Fusarium flocciferum Corda [W&R,G,B]
Syn.: *F. roseum* Lk. emend. Snyd. & Hans. pro parte [S&H]

Perfect state. *Gibberella heterochroma?* Wollenw. (21).
The aerial mycelium is white with a carmine red undersurface. Microconidia are absent. Macroconidia (Fig. 70) are sickle-shaped and similar in shape to *F. sambucinum* but longer and more slender. The basal cell is foot-shaped. Chlamydospores are present and are formed in chains or clusters. The single culture of this species that we have examined is indentical to *F. graminearum*. Furthermore, like *F. graminearum* group 2, it produces *Gibberella* perithecia on CLA within 4 days from a single conidium. These perithecia are indistinguishable from those produced by *G. zeae*.

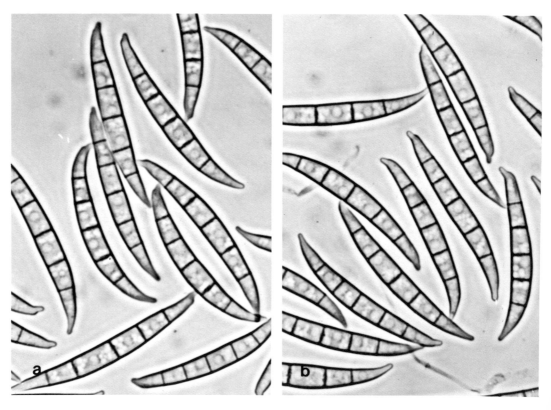

Figure 70. *Fusarium flocciferum:* a, b, macroconidia produced in sporodochia (X1000).

SECTION DISCOLOR

Fusarium sublunatum **Reinking** [W&R,G]

Syn.: *F. tumidum* Sherb? pro parte [B]
F. roseum Lk. emend. Synd. & Hans. pro parte [S&H]

Perfect state. None known.

The aerial mycelium is sparse and white. Sporodochia are orange. Sclerotia are dark blue to blue-green. Microconidia are rare. Macroconidia (Fig. 71) are sickle-shaped with a distinctly foot-shaped basal cell. Chlamydospores are abundant.

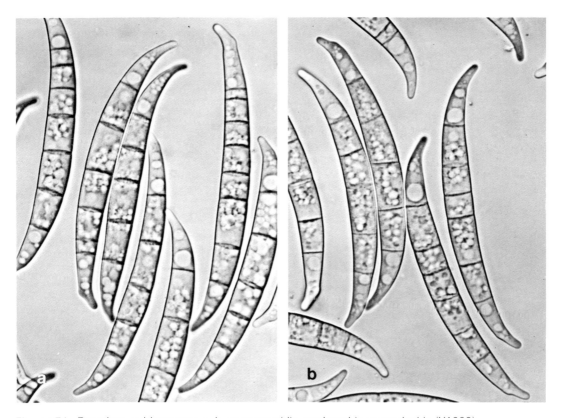

Figure 71. *Fusarium sublunatum:* a, b, macroconidia produced in sporodochia (X1000).

SECTION DISCOLOR
Fusarium tumidum **Sherb.** [W&R,G,B]
Syn.: *F. roseum* Lk. emend. Snyd. & Hans. pro parte [S&H]

Perfect state. None known.

On PDA, the abundant brown sporodochia are present. The undersurface of the colony is also brown. Some microconidia are present. Macroconidia (Fig. 72) are very large and may be 11-septate but are generally 9-septate. The basal cell is notched or has a papilla.

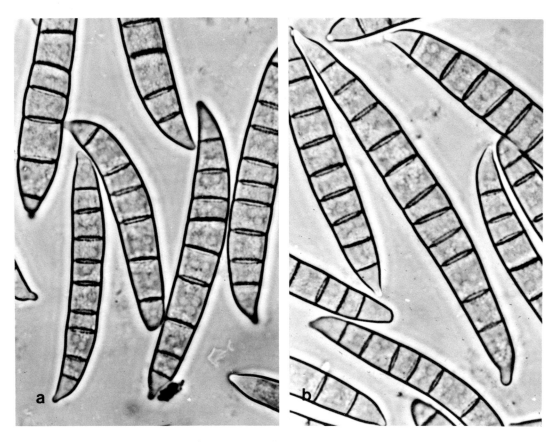

Figure 72. *Fusarium tumidum:* a, b, macroconidia produced in sporodochia (X1000).

SECTION DISCOLOR
Fusarium robustum **Gerlach** (17) [G]

Perfect state. None known.

Growth is rapid, with abundant aerial mycelium and a carmine red undersurface typical of the other members of Wollenweber and Reinking's section Discolor (76). Microconidia are absent. Macroconidia (Fig. 73) resemble those of *F. graminearum* or *F. crookwellense*. Chlamydospores are present.

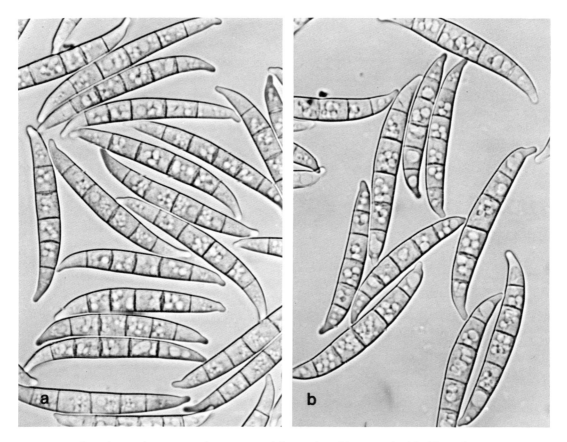

Figure 73. *Fusarium robustum:* a, b, macroconidia produced in sporodochia (X1000).

SECTION DISCOLOR
Fusarium buharicum Jaczewski (21) [G,B]
[Not mentioned by W&R,S&H]

Perfect state. None known.

The single sporodochial culture of this species we have seen produces a colony with white aerial mycelium and dark brown to blue-green sporodochia on PDA. However, the spore masses in pionnotal forms are orange. Microconidia are absent. Macroconidia (Fig. 74) are shaped like the macroconidia of *F. graminearum* with a blunt, sharply bent apical cell. The basal cell is distinctly foot-shaped. Chlamydospores are present. This distinctive species seems to be restricted to southern Russia and northern Iran, as a collar rot pathogen of certain cotton cultivars.

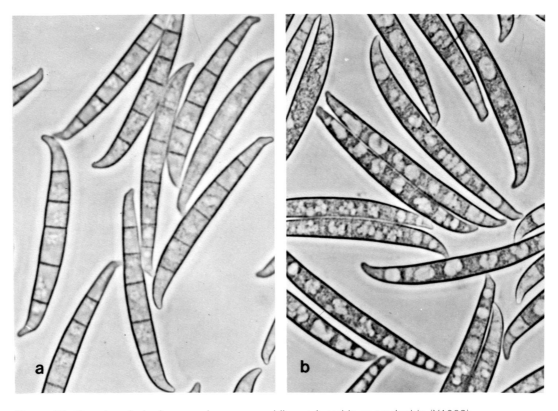

Figure 74. *Fusarium buharicum:* a, b, macroconidia produced in sporodochia (X1000).

SECTION DISCOLOR
Fusarium lunulosporum Gerlach (18) [G]

Perfect state. None known.

On PDA, colonies with scant aerial mycelium and carmine red upper and lower surfaces are produced. Microconidia are absent. Macroconidia (Fig. 75) are characteristically crescent-shaped. The basal cell is distinctly foot-shaped. Chlamydospores are present.

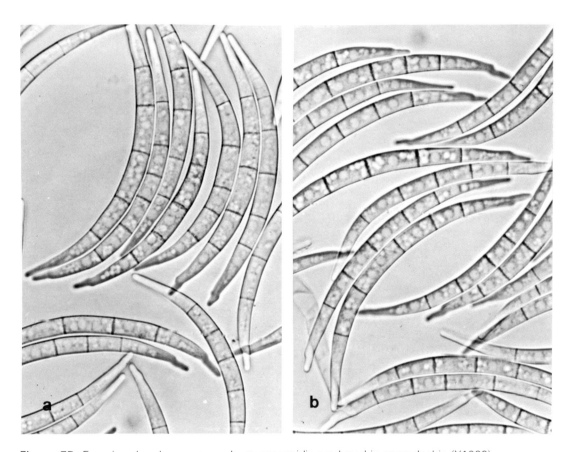

Figure 75. *Fusarium lunulosporum:* a, b, macroconidia produced in sporodochia (X1000).

SECTION LATERITIUM
Fusarium xylarioides Steyaert (4, 21) [B,G]
[Not mentioned by W&R,S&H]

Perfect state. *Gibberella xylarioides* Heim & Saccas (5).
The cultures we have seen on PDA are relatively slow growing, appressed, white to light purple colonies, with little or no aerial mycelium; they are pionnotal in appearance.

Conidia are borne on unbranched and branched monophialides. Microconidia are strongly curved, allantoid, 0–1 septate (Fig. 76a). Macroconidia are strongly curved, 1–3 septate and the basal cell is notched or foot-shaped (Fig. 76a,b). Chlamydospores are rare. Bluish-black protoperithecia develop in some isolates.

The cultures we have seen, including fresh isolates from diseased coffee from Ethiopia, are similar to the "female strain" described by Booth, but we have not seen the so-called "male strain" (4). Von Blittersdorf and Kranz (71) report that the "male strain" (4) differs distinctly from all other isolates of this fungus and most probably is a cultural variant or mutation of *F. stilboides*.

This distinctive species causes "tracheomycosis," a rapid vascular wilt disease of coffee, in the Ivory Coast, Guinea, Central African Republic, Zaire, and Ethiopia.

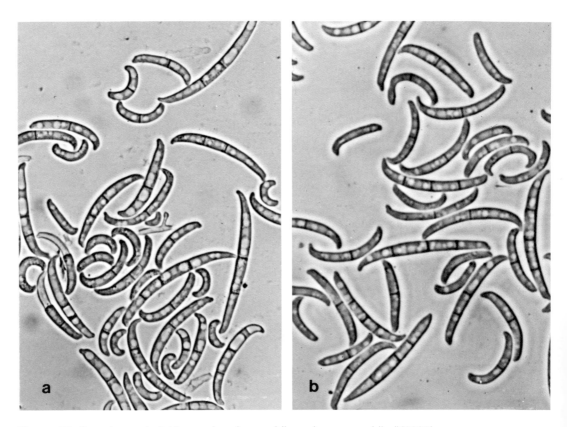

Figure 76. *Fusarium xylarioides:* a, b, microconidia and macroconidia (X1000).

SECTION LISEOLA
Fusarium annulatum Bugnicourt (21, 47) [G]
Syn.: *F. moniliforme* Sheldon pro parte [B]
[Not mentioned by W&R,S&H]

Perfect state. None known.

Cultures are similar to those of the other members of Wollenweber and Reinking's section Liseola (76). Microconidia (Fig. 77), produced in chains and false heads on monophialides and polyphialides, are abundant and are club-shaped with a flattened base. Macroconidia (Fig. 77) are thin-walled, strongly curved and almost ring shaped, with the basal cell distinctly foot-shaped. Chlamydospores are absent. Essentially this species is an *F. proliferatum* type with strongly curved macroconidia.

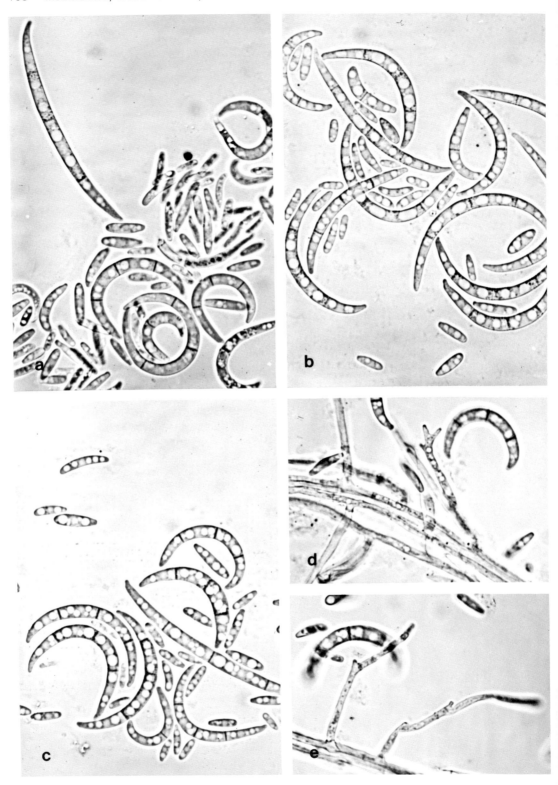

Figure 77. *Fusarium annulatum:* a–c, macroconidia and microconidia; d, e, polyphialides (X1000).

SECTION LISEOLA
Fusarium succisae (Schröter) Sacc. (21, 47) [G]

> Syn.: *F. moniliforme* Sheldon var. *anthophilum* (A. Braun). Wollenw. pro parte
> [W&R]
> *F. moniliforme* Sheldon pro parte [B]
> [Not mentioned by S&H]

Perfect state. None known.

Cultures are similar to those of the other members of Wollenweber and Reinking's section Liseola (76). Microconidia (Fig. 78c) are borne in false heads on monophialides and polyphialides (Fig. 78d,e) and are oval to ellipsoid in shape. Some microconidia have papillae. Macroconidia (Fig. 78a,b) are thin-walled and straight or highly curved almost in a ring on the host and on prepared media such as PDA. The basal cell is distinctly foot-shaped. Chlamydospores are absent. Essentially this species seems to be an *F. subglutinans* type with strongly curved macroconidia. Nirenberg (47) reports that this fungus is abundant on flower heads of *Succisa pratensis* in Europe. We have seen only one isolate and therefore place it in this section with other species of which we have seen only a few cultures.

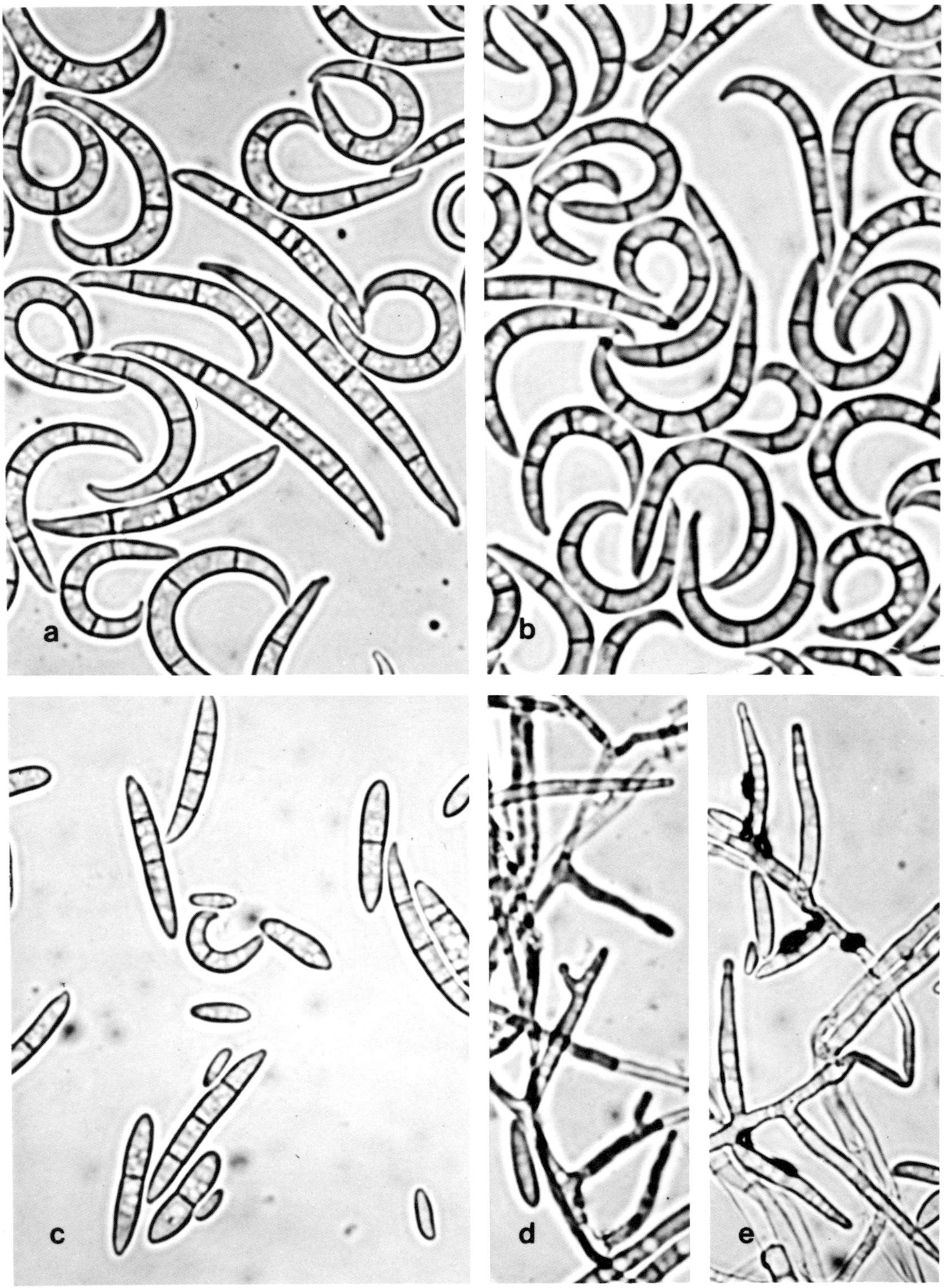

Figure 78. *Fusarium succisae:* a, b, macroconidia produced in sporodochia; c, microconidia produced in the aerial mycelium; d, polyphialide; e, monophialides (X1000).

SECTION ELEGANS or LATERITIUM
Fusarium udum Butler (21) [G,B]
> Syn.: *F. oxysporum* Schlecht. emend. Snyd. & Hans. f. sp. *udum* (Butler) Snyd. &
> Hans. [S&H]
> [Not mentioned by W&R]

Perfect state. *Gibberella indica* B. Rai & R. S. Upadhyay (52).
Microconidia (Fig. 79c, d) are present. Macroconidia (Fig. 79a,b) are present and are
borne on unbranched and branched monophialides. On PDA, the cultures resemble
those of *F. oxysporum.*

Authorities are divided as to the proper place of this fungus and have vacillated between Wollenweber and Reinking's sections Elegans and Lateritium (76).

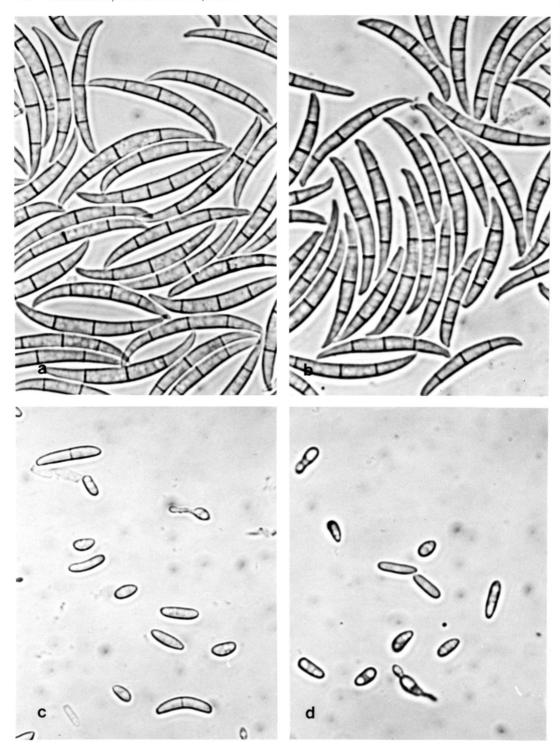

Figure 79. *Fusarium udum:* a, b, macroconidia produced in sporodochia; c, d, microconidia produced in the aerial mycelium (X1000).

List of Cultures Used
in Preparing the Illustrations

The cultures used in preparing the figures for this book have been deposited in the Collection of the Fusarium Research Center, Department of Plant Pathology, The Pennsylvania State University, University Park, Pennsylvania 16802; the South African Medical Research Council Collection, National Research Institute for Nutritional Diseases, P. O. Box 70, Tygerberg 7505, Republic of South Africa; and the Agricultural Research Culture Collection, U.S.D.A., 1815 N. University St., Peoria, Illinois 61604. The identification numbers for the cultures used are listed below for each figure; NA means the culture is not available.

Figure 1—a, E-88; b, d, E-91; c, E-90.

Figure 2—a, b, E-90.

Figure 3—a, E-24; b, E-32; c, E-102; d, E-109.

Figure 4—a, E-3; b, E-7; c, E-24; d, E-9; e, E-52.

Figure 5—a, b, d, e, E-10; c, f, g, E-72.

Figure 6—a, Rd-10; b, c, e, f, g, Rd-35; d, Rd-34.

Figure 7—a, d, N-42; b, N-50; c, N-51.

Figure 8—a, T-412; b, T-558; c, d, T-342.

Figure 9—a, b, T-342.

Figure 10—a, T-388; b, c, d, T-546; e, T-387.

Figure 11—a, b, T-546.

Figure 12—a, c, T-386; b, T-521; d, T-385.

Figure 13—a, T-478; b, T-385; c, d, T-210.

Figure 14—a, T-478; b, T-340; c, T-346; d, e, T-525.

Figure 15—a, T-343; b, T-491; c, T-472; d, T-384.

Figure 16—a, T-548; b, T-491; c, T-472; d, e, f, T-550.

Figure 17—a, NA; b, R-2703.

Figure 18—a, R-2316; b, R-2713; c, d, R-5194

Figure 19—a, R-48; b, R-445; c, R-4954; d, R-5721.

Figure 20—a, c, e, R-4954; b, NA; d, R-4961.

Figure 21—a, R-773; b, R-836; c, R-2138; d, R-2182.

Figure 22—a, e, f, R-2181; b, R-4333; c, d, R-4237; g, NA.

Figure 23—a, b, d, R-5201; c, e, R-5200.

Figure 24—a, R-454; b, R-2721; c, R-2711; d, NA.

Figure 25—a, R-1011; b, R-920; c, R-4306; d, R-908.

Figure 26—a, b, R-6327; c, R-6507; d, e, f, R-6720.

Figure 27—a, R-2224; b, R-3600; c, R-3625; d, R-5137.

Figure 28—a, b, R-5137; c, d, R-3468; e, f, g, R-2227.

Figure 29—a, R-2121; b, R-6363; c, R-6253; d, R-6250.

Figure 30—a, b, e, R-2121; c, d, R-6279; f, g, R-6230.

Figure 31—a, R-2109; b, R-2136; c, R-2636; d, R-2641.

Figure 32—a, R-2238; b, R-3618; c, R-5854; d, R-6365.

Figure 33—a, R-3120; b, c, d, R-6202; e, f, R-6638.

Figure 34—a, R-4421; b, R-314; c, d, e, R-4422.

Figure 35—a, R-5128; b, R-4422.

Figure 36—a, b, R-5195; c, d, R-6669.

Figure 37—a, R-3124; b, R-3052; c, R-382; d, e, R-6479.

Figure 38—a, R-583; b, R-2155; c, R-2633; d, R-2882.

Figure 39—a, R-5753; b, R-5683; c, R-5690; d, R-5684; e, R-5344.

Figure 40—a, R-358; b, R-444; c, R-699; d, R-5063.

Figure 41—a, b, c, R-1274.

Figure 42—a, R-669; b, R-2581; c, NA; d, R-2808.

Figure 43—a, NA; b, c, R-6733.

Figure 44—a, R-2770; b, R-3933; c, R-2201; d, R-3090.

Figure 45—a, b, R-3090; c, R-2153; d, R-3611; e, R-3582.

Figure 46—a, L-47; b, L-66; c, L-82; d, L-103.

Figure 47—a, L-107; b, c, d, L-84; e, L-88.

Figure 48—a, L-108; b, L-47; c, d, L-80; e, L-105.

Figure 49—a, c, d, M-1142; b, M-1213.

Figure 50—a, b, c, e, M-1213; d, M-1068.

Figure 51—a, b, c, M-1217; d, M-944.

Figure 52—a, M-1219; b, d, e, M-944; c, M-1264.

Figure 53—a, d, M-848; b, c, M-851.

Figure 54—a, M-850; b, M-851; c, M-1168; d, e, M-1138.

Figure 55—a, M-1211; b, M-1190; c, M-843; d, M-1238.

Figure 56—a, M-1211; b, M-1090; c, M-843; d, e, f, M-1238.

Figure 57—a, O-45; b, O-1153; c, O-1155; d, O-681.

Figure 58—a, O-1142; b, O-1146; c, d, e, O-1141.

Figure 59—a, b, c, e, O-1080; d, O-1154; f, g, O-1155.

Figure 60—a, S-70; b, S-86; c, NA; d, S-145.

Figure 61—a, S-706; b, S-714; c, d, NA.

Figure 62—a, S-401; b, S-706; c, S-415, d, S-128; e, S-203.

Figure 63—a, S-714; b, S-688; c, S-655; d, e, S-189.

Figure 64—a, E-94; b, E-93.

Figure 65—a, b, E-98.

Figure 66—a, b, c, d, E-99.

Figure 67—a, b, N-40.

Figure 68—a, b, R-5199.

Figure 69—a, R-5198; b, c, R-5197.

Figure 70—a, b, R-5204.

Figure 71—a, b, R-5213.

Figure 72—a, b, R-5823.

Figure 73—a, R-5821; b, R-5700.

Figure 74—a, b, R-4955.

Figure 75—a, b, R-5822.

Figure 76—a, b, L-101.

Figure 77—a, b, c, d, e, M-1220.

Figure 78—a, b, c, d, e, M-1221.

Figure 79—a, c, d, O-1116; b, O-1117.

Part IV
Taxonomic Systems and Perfect States

Taxonomic Systems
for *Fusarium* Species

There is no single taxonomic system in use today that is completely satisfactory for the identification of all *Fusarium* species. The continued proliferation of "new" or "modern" systems for the taxonomy of *Fusarium* species will not solve the problem. In fact such activity only adds to the current state of confusion regarding *Fusarium* taxonomy. These statements give the reader some indication of the authors' current approach to the taxonomy of *Fusarium* species. After more than 20 years of research and study on this problem, our conclusion is that there is enough good information within the existing systems for identification. The existing taxonomic systems for identification of *Fusarium* species have been discussed (67). If the existing information plus some new information is combined and properly utilized, a usable, practical system for the identification of *Fusarium* species can be compiled. The system presented in this volume takes the publication of Wollenweber and Reinking (76) as the starting point for Fusarium taxonomy and combines the information from several taxonomic systems, developed from 1935 to the present (2, 4, 21, 24, 35, 38, 41, 54, 58, 59, 60, 76), with results from our own research on Fusarium taxonomy.

In 1935, Wollenweber and Reinking (76) published their monumental work on Fusarium taxonomy. This volume is well illustrated with very accurate line drawings made by Wollenweber. In fact no one has succeeded in making line drawings of the *Fusarium* species of comparable quality since then. Wollenweber also had other publications (73,

74, 75) relating to this taxonomic system, but the book by Wollenweber and Reinking (76), culminating 40 years of research, has become the standard reference work on this subject. In fact, all systems for Fusarium taxonomy are based on this publication. Wollenweber and Reinking began with approximately 1,000 named species of *Fusarium* and organized these into 16 sections containing 65 species, 55 varieties, and 22 forms. One can begin to appreciate the magnitude of this task by considering that Wollenweber and Reinking list 77 synonyms for *F. avenaceum* alone, and 133 synonyms for *F. lateritium* and its perfect state.

A distinct set of characteristics was used to separate each group. The characteristics used to separate sections were i) the presence or absence of microconidia, ii) the shape of the microconidia, iii) the presence or absence of chlamydospores, iv) the location of the chlamydospores—intercalary or terminal, v) the shape of the macroconidia, and vi) the shape of the basal or foot cells on the macroconidia. The 16 sections of Wollenweber and Reinking are listed in Table 1.

The sections were divided into species, varieties, and forms on the basis of i) the color of the stroma, ii) the presence or absence of sclerotia, iii) the number of septations in the macroconidia, and iv) the length and width of the macroconidia. For instance, in the section Elegans (Table 2) great emphasis was placed on measurements of the length and width of macroconidia; species, varieties, and forms were separated on the basis of a few microns difference in length or width and on the number of septations in the macroconidia. Each isolate studied was grown on six different media: beerwort agar, carrot decoction agar, oatmeal agar, rice mash, alfalfa stems, and barley heads. In some cases potato dextrose agar and potato pieces were also used. Observation of cultures grown on these media tends to emphasize differences rather than similarities, and to exaggerate minor differences, such as length and width of macroconidia, which results in finer and finer separations at the species, variety, and form level. This produces a system so complex that it is difficult or impossible to use it to construct a satisfactory practical key. The characteristics used by Wollenweber and Reinking (76) to separate species, varieties, and forms are not stable and can be altered readily by growing cultures on various media and under various environmental conditions.

Two other problems may also be partly responsible for the complexity of this system. One is that we are not sure that the problem of cultural variation or mutation in *Fusarium* was recognized by Wollenweber and Reinking (76). If it was not, then, given that we know their cultures were not started from single spores, it is likely that a few of their species and many of their varieities and forms are nothing more than cultural mutants of *Fusarium* species. Some examples may be *F. sublunatum* and *F. sublunatum* var. *elongatum*, *F. trichothecioides*, *F. merismoides* var. *crassum*, *F. dimerum* var. *nectriodes* and *F. dimerum* var. *violaceum*. The other problem is that some species may have been named on the basis of only one or two cultures. We believe one must examine a population of the organism in question in order to determine the range of variation that may occur in any given entity. Looking at only one culture of an entity cannot reveal that range; it can, however, produce considerable confusion and difficulty for others.

Figure 80 shows the relationship of several other taxonomic systems for the identification of *Fusarium* species to that of Wollenweber and Reinking (76) and to each other. Fusarium taxonomists have often been divided into "lumpers" and "splitters." In this figure the "splitters" are in the middle with the "lumpers" on the right and the "moderate splitters" on the left. Gerlach (16, 20, 21, 22) continues to work in Wollenweber's laboratory at the Biologische Bundesanstalt, West Berlin. Both his philosophy and the techniques he uses in studying *Fusarium* and establishing new species place him decid-

Table 1. The relationship of the sixteen sections of Wollenweber and Reinking to the nine species of Snyder and Hansen.

Sections of Wollenweber and Reinking	Species of Snyder and Hansen
Eupionnotes	*episphaeria*
Macroconia	*episphaeria*
Spicarioides	*rigidiuscula*
Submicrocera	*none*
Pseudomicrocera	*none*
Arachnites	*nivale*
Sporotrichiella	*tricinctum*
Roseum	*roseum*
Arthrosporiella	*roseum*
Gibbosum	*roseum*
Discolor	*roseum*
Lateritium	*lateritium*
Liseola	*moniliforme*
Elegans	*oxysporum*
Martiella	*solani*
Ventricosum	*solani*

Table 2. Characteristics of the three subsections of the section Elegans (10 species, 18 varieties, 12 forms).

Subsection Orthocera (5 species, 6 varieties, 1 form)
  Sporodochia absent
  Macroconidia 3 to 5 septate
  Macroconidia 3–4 μm × 27–50 μm

Subsection Constrictum (1 species, 5 varieties)
  Sporodochia present
  Macroconidia 3 to 5 septate
  Macroconidia 3–3.7 μm × 30–55.5 μm

Subsection Oxysporum (4 species, 7 varieties, 11 forms)
  Sporodochia present
  Macroconidia 3 to 5 septate
  Macroconidia 3.7–5 μm × 32—47 μm

edly with the "splitters." This is evident from the 90+ species that appear in his new atlas (22), a well-illustrated work that uses excellent photographs and line drawings to supplement Wollenweber's original drawings, which are also used in part. Gerlach continues to grow cultures on the eight different media used by Wollenweber and Reinking (76) and under conditions that accentuate differences. He concentrates on these differences rather than on similarities, with the result that a slight cultural difference may be the basis for a new species (17, 18, 55) or variety (19). New species are established based on a single culture (17, 18, 55) and in some cases on a single mutant culture (17). This philosophy leads to a complex taxonomic system that is difficult to use for the same reasons that Wollenweber and Reinking's system is difficult to use.

The systems of Raillo (53, 54) and Bilai (2, 3) are not as well known as the other systems shown in Figure 80. In one of Raillo's papers (53) she studied morphological characters useful in taxonomy. From this study she concluded that i) the form of the top or apical cell is the guiding character in species determination; ii) the incurvature of conidia, length of the top or apical cell, number of septa, and width of conidia are the characters used in separating subspecies and varieties; and iii) cultural characters such as pigment, presence of sclerotia, and mode of spore formation are characters used in separating forms only. She also studied variability in *Fusarium* by use of the single-spore technique. She found that i) the form of the top cell and the incurvature of conidia remain constant in isolates developed from a single spore; ii) the number of septa is constant in isolates within a single-spore culture; iii) the length and width of conidia vary considerably in separate isolates within a single-spore culture; iv) the number of sclerotia varies greatly in separate isolates within a single-spore culture; iv) the number of sclerotia varies greatly in separate isolates within a single-spore culture; and v) the mode of spore formation (pionnotes, pseudo-pionnotes, and sporodochia) varies in separate isolates within a single-spore culture.

Bilai (2, 3), a Russian researcher, recognized the importance of cultural variation or mutation in Fusarium taxonomy. She did a critical analysis of several characteristics used in Fusarium taxonomy by studying experimental variability of individual isolates and establishing the range of variation for some species (3). In addition she studied experimental morphogenesis in single-spore isolates in culture, paying particular attention to the effects of temperature, moisture, length of growth period, and composition of the medium as well as the method of germination and aging of conidia. Her results showed that the range of variability was greater than that accepted in the diagnosis of many species and often included the features of the whole section. On the basis of these results she revised the taxonomy of the genus to include only 9 sections, 26 species, and 29 varieties. Some of her changes, such as combining the section Liseola with the section Elegans and combining the section Gibbosum with section Discolor, are difficult to understand. Although this system may have been used in Russia, it has not been accepted and used in other parts of the world.

Joffe (35) began working on *Fusarium* in the late 1940s in Russia. Later he emigrated to Israel and continued his work. He examined a large number of isolates of *Fusarium* from soil, wilting or decaying plants, and seed. These isolates were collected in the warm, semi-arid climate of Israel and the cold climate of the U.S.S.R. Other isolates were received from research institutes in several countries. His philosophy and approach to Fusarium taxonomy is similar to that of Wollenweber and Reinking (76) and Gerlach (16, 20, 21, 22). In fact his so-called "modern system" appears to be simply a restatement of Wollenweber and Reinking's sections and species with the addition of some names by Gerlach (21, 22). He recognizes 13 sections, 33 species, and 14 varieties.

Gordon worked in Canada from the 1930s to the 1960s. He published a series of papers (23, 24, 25, 26, 27, 28, 29, 30) detailing his taxonomic system. The paper published in 1952 (24) is the most important and gives a detailed description of the system he proposed. He worked with *Fusarium* species isolated from cereal seed, various host plants, and soil from both temperate and tropical geographic areas. His system of taxonomy is closer to that of Wollenweber and Reinking (76) than to that of Snyder and Hansen (58, 59, 60). Certain sections of the genus, particularly Lateritium, Liseola, Elegans, and Martiella, were modified by the adoption of the revisions of these sections, as a whole or in part, that were proposed by Snyder and Hansen (58, 59, 60). Therefore, his taxonomic system was a compromise between that of Wollenweber and Reinking (76) and that of Snyder and Hansen (58, 59, 60).

Booth (4) modified Gordon's system and added information based on his studies. He expanded the information on the perfect states. A major contribution was the addition of the information on conidiophores and conidiogenous cells that is useful in the taxonomy of *Fusarium* species. He pointed out the value of the presence of polyphialides versus monophialides in separating sections and species. The length and shape of the microconidiophores was also shown to be a reliable character to use in separating *F. oxysporum, F. solani,* and *F. moniliforme.* Booth made a real effort to bridge the gap between the taxonomic mycologists and plant pathologists and other groups that must work with these organisms on a regular basis.

Snyder and Hansen are considered to be the ultimate "lumpers." In the 1930s W. C. Snyder went to Berlin and spent a year working with Wollenweber in his laboratory. When he returned to Berkeley he began an extensive research program on the biology and taxonomy of *Fusarium* species in cooperation with H. N. Hansen, who pioneered the use of single-spore cultures. In the 1940s they published their ideas on the taxonomy of *Fusarium* species in three papers (58, 59, 60). In essence this system makes nine species out of Wollenweber and Reinking's 16 sections (Table 1). Snyder and Hansen's system is based primarily on the morphology of the macroconidia and an extensive study of the general nature and variability of *Fusarium* species. The basis for their work was an extensive single-spore analysis of cultures of *Fusarium* species under identical conditions of substrate and other environmental conditions. These studies revealed a great range of variability in spore length, width, and septation, in kinds and intensities of pigments produced, and in the presence or absence of sporodochia and sclerotia among the subcultures of the same original single-spore culture. In their work with members of Wollenweber and Reinking's section Elegans (76) they found that progeny of a single parent may be placed in different species and even in different sub-sections. This is an indication that the characters used for speciation by Wollenweber and Reinking (76) were too narrow.

Snyder and Hansen's work with *F. oxysporum* (58) (section Elegans) is the basis for their system. This work illustrated the importance of cultural variation in taxonomy and is excellent work generally accepted by most people interested in Fusarium taxonomy. Their work with *F. solani* (59), which is also generally accepted, showed that the variations are inheritable. The remaining work (60), including the lumping of several sections into one species, is not generally accepted. For example, the lumping of Wollenweber and Reinking's sections Arthrosporiella, Discolor, Gibbosum, and Roseum (76) into *F. roseum* has caused a great deal of confusion and controversy. The great reduction of species eliminated the convenience of naming certain fungi which had previously been known as species. In order to overcome this objection Snyder, Hansen, and Oswald proposed the use of nonbotanical, horticultural variety names or cultivars (65) and additional cultivars were named later (44). Because the cultivar was to be an informal device, no formal guidelines for their delineation were laid down, nor was a definitive listing made. The cultivar system has been difficult to use because no accurate descriptions of the various cultivars are available. Another difficult area involved the lumping of all species in section Sporotrichiella into the single species *F. tricinctum.* This is especially confusing with regard to the mycotoxin-producing strains. This is not satisfactory and we recognize the original four species of Wollenweber and Reinking (76), namely *F. poae, F. tricinctum, F. sporotrichioides,* and *F. chlamydosporum.*

Messiaen (40) and Messiaen and Cassini (41) essentially follow Snyder and Hansen's system, with some modifications. The major modification they made was to use botanical varieties instead of cultivars at the subspecies level in *F. roseum.* They provide descriptions for each variety. A key is also provided for the entire system.

Matuo (38) also follows the Snyder and Hansen system and provides a key to the entire system. Matuo and Kobayashi (39) reported that *Hypocrea splendens* produced a conidial state which they named *F. splendens.* However further work showed that this was most likely a *Nectria* hyperparasite (5). Matuo is also in favor of lumping *F. lateritium* and *F. roseum,* but this concept has received very little support.

Faced with this array of differing systems and opinions regarding Fusarium taxonomy, how does one make sense out of the existing information and use it in the best way possible for the identification of *Fusarium* species? We have selected what we consider to be the best parts of each system and combined them along with results of our own research to develop a compromise system in which utility for practical identification is emphasized. We have used *F. oxysporum* and *F. solani* as described by Snyder and Hansen (58, 59). The information on conidiophores and conidiogenous cells, especially that on microconidiophores as supplied by Booth (3), has been incorporated. Those of Wollenweber and Reinking's sections (76) that contain important toxigenic species, namely Sporotrichiella, Liseola, Roseum, Gibbosum, Discolor, and Arthrosporiella, have been retained. Other sections from Wollenweber and Reinking (76) that have been maintained are Eupionnotes, Macroconia, Spicarioides, and Lateritium. In all cases the number of species in these sections has been reduced and the varieties and forms have been combined with the appropriate species. We are of the opinion that many of the varieties and forms may be cultural variants. In this way we are presenting a system which we believe is usable for practical identification.

Figure 80. Principal taxonomic systems for *Fusarium* species.

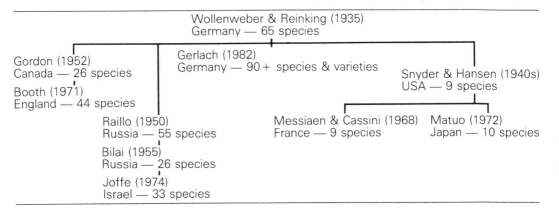

Perfect States

The perfect states of *Fusarium* species may be unfamiliar to many people working with these fungi. Plant pathologists most often deal with the imperfect state, and the perfect states may have little to do with the disease problem under study. Booth (5) notes that the presence of the perfect or porithocial state in *Fusarium* appears, superficially at least, to be of more value to the taxonomist than to the fungus. In fact, some of the most successful *Fusarium* species, such as *F. oxysporum* and *F. culmorum,* appear to have lost their sexual ability and to have adopted other methods of facilitating genetic adaptions (5).

The perfect states of *Fusarium* species all belong in the order Hypocreales, subclass Ascomycotina, class Euascomycetes or Pyrenomycetes (4). This order is characterized by a definite perthecial wall which is soft, fleshy or membranous, and usually bright colored. The perithecia have a definite ostiole. The ascospores are usually 1–3 septate. The genera *Nectria, Gibberella,* and *Calonectria* have *Fusarium*-type conidial states. In these genera the perithecia are all superficial, with or without a stroma. *Calonectria* is separated from *Nectria* because it has two or more transverse septa in the ascospores, and from *Gibberella* because it lacks the purple pigment generally present in *Gibberella.* Booth (4) used *Micronectriella* for the perithecial forms of *F. nivale, F. tabacinum,* and *F. stoveri.* Some question exists as to the status of this genus in the Hypocreales, and in a later publication he reported that since the name

Micronectriella is now untenable, the name *Plectosphaerella* should be used (5). Gerlach (21) also uses the name *Plectosphaerella.*

Wollenweber and Reinking (76) (Table 3) recognize *Nectria, Calonectria, Gibberella,* and *Hypomyces* as perfect states of *Fusarium* species. This concept is followed by Snyder and Toussoun (64) and Messiaen and Cassini (41).

Wollenweber (73) placed all the nectroid species with chlamydospores in *Hypomyces.* Booth (5) points out that this is not a valid separation because the presence or absence of chlamydospores in ascospore cultures of *Nectria* is not a generic distinction. Therefore *Nectria* is the correct genus for these forms with *Fusarium* imperfect states.

Gerlach (21) (Table 3) recognizes *Nectria, Calonectria,* and *Gibberella* as perfect states of *Fusarium* species. In addition to these he uses *Plectosphaerella* as the perfect state for some of the species in the section Eupionnotes and *Monographella* for some of the species in the section Arachnites.

Booth (5) (Table 3) recognizes *Nectria, Calonectria,* and *Gibberella* as perfect states of *Fusarium* species. He also uses *Plectosphaerella* and *Monographella* as perfect states for species in his section Arachnites. The main area of contention seems to center on the correct name for the perfect state of *F. nivale.* This subject is discussed in detail by Booth (5) in a recent publication.

Table 3. Perfect states of *Fusarium* species recognized by Wollenweber & Reinking (76), Gerlach (21), and Booth (5).

Sections of Wollenweber and Reinking	Perfect States		Sections of Booth	Perfect States of Booth
	Wollenweber and Reinking	Gerlach		
Eupionnotes	*Nectria*	*Nectria, Plectosphaerella*	Episphaeria	*Nectria*
Macroconia	*Nectria*	*Nectria*	Episphaera	*Nectria*
			Coccophilum	*Nectria, Calonectria*
Spicarioides	*Calonectria*	*Calonectria*	Spicarioides	*Calonectria*
Submicrocera	*Calonectria*	*Calonectria*	Arachnites	*Monographella, Plectosphaerella*
Pseudomicrocera	*Calonectria*	*Calonectria*	Coccophilum	*Nectria, Calonectria*
Arachnites	*Calonectria*	*Calonectria, Monographella*	Arachnites	*Monographella, Plectosphaerella*
Sporotrichiella	*None*	*None*	Sporotrichiella	*None*
Roseum	*None*	*Gibberella*	Arthrosporiella	*Gibberella*
Arthrosporiella	*None*	*None*	Arthrosporiella	*Gibberella*
Gibbosum	*Gibberella*	*Gibberella*	Arthrosporiella	*Gibberella*
Discolor	*Gibberella*	*Gibberella*	Gibbosum	*Gibberella*
Lateritium	*Gibberella*	*Gibberella*	Discolor	*Gibberella*
Liseola	*Gibberella*	*Gibberella*	Lateritium	*Gibberella*
Elegans	*None*	*None*	Liseola	*Gibberella*
Martiella	*Hypomyces*	*Nectria*	Elegans	*None*
Ventricosum	*Hypomyces*	*Nectria*	Martiella	*Nectria*
			Martiella	*Nectria*

Literature Cited

1. Alexander, J. V., J. A. Bourret, A. H. Gold, and W. C. Snyder. 1966. Induction of chlamydo-spore formation by *Fusarium solani* in sterile soil extracts. Phytopathology 56:353–54.
2. Bilai, V. I. 1955. The Fusaria (Biology and Systematics). Kiev: Akad. Nauk. Ukr. SSR. 320 pp.
3. Bilai, V. I. 1970. Experimental morphogenesis in the fungi of the genus *Fusarium* and their taxonomy. Ann. Acad. Sci. Fenn. A, IV Biologica 168: 7–18.
4. Booth, C. 1971. The Genus *Fusarium*. Commonwealth Mycol. Inst., Kew, England. 237 pp.
5. Booth, C. 1981. Perfect states (teleomorphs) of *Fusarium* species, pp. 446–52. *In* P. E. Nelson, T. A. Toussoun, and R. J. Cook (eds.), *Fusarium:* Diseases, Biology, and Taxonomy. University Park: Pennsylvania State Univ. Press. 457 pp.
6. Burgess, L. W., P. E. Nelson, and T. A. Toussoun. 1981. Laboratory Manual for Fusarium Studies. Prepared for Fusarium Workshop, Univ. Sydney. May 1981. 50 pp.
7. Burgess, L. W., P. E. Nelson, and T. A. Toussoun. 1982. Characterization, geographic distribution, and ecology of *Fusarium crookwellense* sp. nov. Trans. Brit. Mycol. Soc. 79:497–505.
8. Burgess, L. W., P. E. Nelson, T. A. Toussoun, and W. F. O. Marasas. 1983. *Fusarium scirpi:* Emended description and notes on geographic distribution. (In Preparation).
9. Burgess, L. W., H. J. Ogle, J. P. Edgerton, L. L. Stubbs, and P. E. Nelson. 1973. The biology of fungi associated with root rot of subterranean clover in Victoria. Proc. Royal Soc. Vict. 86 (Part I): 19–28.
10. Burgess, L. W., A. H. Wearing, and T. A. Toussoun. 1975. Surveys of Fusaria associated with crown rot of wheat in eastern Australia. Australian J. Agric. Res. 26:791–99.

11. Cook, R. J. 1967. *Gibberella avenacea* sp. n., perfect stage of *Fusarium roseum* f. sp. *cerealis* 'Avenaceum'. Phytopathology 57:732–36.

12. El-Gholl, N. E., J. J. McRitchie, C. L. Schoulties, and A. H. Ridings. 1978. The identification, induction of perithecia, and pathogenicity of *Gibberella (Fusarium) tricincta* n. sp. Can. J. Bot. 56:2203–6.

13. Fisher, N. L., L. W. Burgess, T. A. Toussoun, and P. E. Nelson. 1982. Carnation leaves as a substrate and for preserving cultures of *Fusarium* species. Phytopathology 72:151–53.

14. Fisher, N. L., W. F. O. Marasas, and T. A. Toussoun. 1983. Taxonomic importance of microconidial chains in *Fusarium* section Liseola and effects of water potential on their generation. Mycologia 75: (In Press).

15. Gams, W., and E. Müller. 1980. Conidiogenesis of *Fusarium nivale* and *Rhynchosporium oryzae* and its taxonomic implications. Neth. J. Plant Pathol. 86:45–53.

16. Gerlach, W. 1970. Suggestions to an acceptable modern *Fusarium* system. Ann. Acad. Sci. Fenn. A, IV Biologica 168:37–49.

17. Gerlach, W. 1977. *Fusarium robustum* spec. nov. der Erreger einer Stammfaule an *Araucaria angustifolia* (Bertol.) O. Kuntz in Argentinien? Phytopathol. Z. 88:29–37.

18. Gerlach, W. 1977. *Fusarium lunulosporum* spec. nov. von Grapefruit aus Südafrika, ein Fruchtfäulerreger. Phytopathol. Z. 88:280–84.

19. Gerlach, W. 1977. Drei neue Varietäten von *Fusarium merismoides, F. larvarum* und *F. chlamydosporum.* Phytopathol. Z. 90:31–42.

20. Gerlach, W. 1978. Critical remarks on the present situation in Fusarium taxonomy, pp. 115–24. *In* C. V. Subramanian (ed.), Taxonomy of Fungi. Proc. Int. Symp. Taxonomy Fungi. Univ. Madras. Part 1. 304 pp.

21. Gerlach, W. 1981. The present concept of *Fusarium* classification, pp. 413–26. *In* P. E. Nelson, T. A. Toussoun, and R. J. Cook (eds.), *Fusarium:*Diseases, Biology, and Taxonomy. University Park: Pennsylvania State Univ. Press. 457 pp.

22. Gerlach, W., and H. Nirenberg. 1982. The genus *Fusarium* — A Pictorial Atlas. Mitt. Biol. Bundesanst. Land-Forstwirtsch. Berlin-Dahlem 209:1–406.

23. Gordon, W. L. 1944. The occurrence of *Fusarium* species in Canada. I. Species of *Fusarium* isolated from farm samples of cereal seed in Manitoba. Can. J. Res., C, 22:282–86.

24. Gordon, W. L. 1952. The occurrence of *Fusarium* species in Canada. II. Prevalence and taxonomy of *Fusarium* species in cereal seed. Can. J. Bot. 30:209–51.

25. Gordon, W. L. 1954. The occurrence of *Fusarium* species in Canada. III. Taxonomy of *Fusarium* species in the seed of vegetable, forage, and miscellaneous crops. Can. J. Bot. 32:576–90.

26. Gordon, W. L. 1954. The occurrence of *Fusarium* species in Canada. IV. Taxonomy and prevalence of *Fusarium* species in the soil of cereal plots. Can. J. Bot. 32:622–29.

27. Gordon, W. L. 1956. The occurrence of *Fusarium* species in Canada. V. Taxonomy and geographic distribution of *Fusarium* species in soil. Can. J. Bot. 34:833–46.

28. Gordon, W. L. 1956. The taxonomy and habitats of the *Fusarium* species in Trinidad, B. W. I. Can. J. Bot. 34:847–64.

29. Gordon, W. L. 1959. The occurrence of *Fusarium* species in Canada. VI. Taxonomy and geographic distribution of *Fusarium* species on plants, insects, and fungi. Can. J. Bot. 37:257–90.

30. Gordon, W. L. 1960. The taxonomy and habitats of *Fusarium* species from tropical and temperate regions. Can. J. Bot. 38:643–58.

31. Hansen, H. N., and W. C. Snyder. 1947. Gaseous sterilization of biological materials for use as culture media. Phytopathology 37:369–71.

32. Hatt, H. (ed.). 1980. American type culture collection methods. I. Laboratory manual on preservation freezing and freeze-drying. Amer. Type Culture Collection, Rockville, Maryland. 51 pp.

33. Horst, R. K., P. E. Nelson, and T. A. Toussoun. 1970. Aerobiology of *Fusarium* spp. associated with stem rot of *Dianthus caryophyllus.* Phytopathology 60:1296 (Abstr.).

34. Hwang, Shuh-Wei. 1948. Variability and perithecium production of a homothallic form of the fungus, *Hypomyces solani*. Farlowia 3:315–26.

35. Joffe, A. Z. 1974. A modern system of Fusarium taxonomy. Mycopathol. Mycol. Appl. 53:201–28.

36. Komada, H. 1975. Development of a selective medium for quantitative isolation of *Fusarium oxysporum* from natural soil. Rev. Plant Prot. Res. 8:114–25.

37. Lukezic, F. L., and W. J. Kaiser. 1966. Aerobiology of *Fusarium roseum* 'Gibbosum' associated with crown rot of boxed bananas. Phytopathology 56:545–48.

38. Matuo, T. 1972. Taxonomic studies of phytopathgenic Fusaria in Japan. Rev. Plant Prot. Res. 5:34–45.

39. Matuo, T., and T. Kobayashi. 1960. A new *Fusarium,* the conidial state of *Hypocrea splendens* Phil. & Plowr. Trans. Mycol. Soc. Japan 2:13–15.

40. Messiaen, C. M. 1959. La systématique du genre *Fusarium* selon Snyder et Hansen. Rev. Pathol. Veg. Entomol. Agric. Fr. 38:253–66.

41. Messiaen, C. M., and R. Cassini. 1968. Recherches sur les fusarioses. IV. La systématique des *Fusarium*. Ann. Epiphyt. 19:387–454.

42. Meyer, E. 1955. The preservation of dermatophytes at sub-freezing temperatures. Mycologia 47:664–68.

43. Nash, S. M., and W. C. Snyder. 1962. Quantitative estimations by plate counts of propagules of the bean root rot *Fusarium* in field soils. Phytophatology 52:567–72.

44. Nash, S. M., and W. C. Snyder. 1965. Quantitative and qualitative comparisons of *Fusarium* populations in cultivated fields and noncultivated parent soils. Can. J. Bot. 43:939–45.

45. Neish, G. A., and M. Leggett. 1981. *Fusarium moniliforme* var. *intermedium,* a new variety in the Liseola section. Can. J. Bot. 59:288–91.

46. Nelson, P. E., B. W. Pennypacker, T. A. Toussoun, and R. K. Horst. 1975. Fusarium stub dieback of carnation. Phytopathology 65:575–81.

47. Nirenberg, H. 1976. Untersuchungen über die morphologische und biologische Differenzierung in der Fusarium-Sektion Liseola. Mitt. Biol. Bundesanst Land-Forstwirtsch. Berlin-Dahlem 169:1–117.

48. Nirenberg, H. I. 1981. A simplified method for identifying *Fusarium* spp. occurring on wheat. Can. J. Bot. 59:1599–1609.

49. Ooka, J. J., and T. Kommedahl. 1977. Wind and rain dispersal of *Fusarium moniliforme* in corn fields. Phytopathology 67:1023–26.

50. Oswald, J. W. 1949. Cultural variation, taxonomy and pathogenicity of *Fusarium* species associated with cereal root rots. Phytopathology 39:359–76.

51. Prasad, N. 1949. Variability of the cucurbit root-rot fungus, *Fusarium* (*Hypomyces*) *solani* f. *cucurbitae*. Phytopathology 39:133–41.

52. Rai, B., and R. S. Upadhyay. 1982. *Gibberella indica:* The perfect state of *Fusarium udum*. Mycologia 74:343–46.

53. Raillo, A. 1935. Diagnostic estimation of morphological and cultural characters in the genus *Fusarium*. Bull. Plant Prot. II., Leningrad (Phytopathol.) 7:1–100.

54. Raillo, A. 1950. Griby roda *Fuzarium*. State Publ. Moskva: Gos. izd-vo selk-hoz. lit-ry. 415 pp.

55. Schneider, R., and J. Dalchow. 1975. *Fusarium inflexum* spec. nov., als Erreger einer Welke-krankheit an *Vicia faba* L. in Deutschland. Phytopathol. Z. 82:70–82.

56. Snyder, W. C. 1940. White perithecia and the taxonomy of *Hypomyces ipomoeae*. Mycologia 32:646–48.

57. Snyder, W. C., and H. N. Hansen. 1939. The importance of variation in the taxonomy of fungi. Proc. 6th Pacific Sci. Congr. 4:749–52.

58. Snyder, W. C., and H. N. Hansen. 1940. The species concept in *Fusarium*. Amer. J. Bot. 27:64–67.

59. Snyder, W. C., and H. N. Hansen. 1941. The species concept in *Fusarium* with reference to section Martiella. Amer. J. Bot. 28:738–42.

60. Snyder, W. C., and H. N. Hansen. 1945. The species concept in *Fusarium* with reference to Discolor and other sections. Amer. J. Bot. 32:657–66.
61. Snyder, W. C., and H. N. Hansen. 1946. Control of culture mites by cigarette paper barriers. Mycologia 38:455–62.
62. Snyder, W. C., and H. N. Hansen. 1947. Advantages of natural media and environments in the culture of fungi. Phytopathology 37:420–21.
63. Snyder, W. C., and H. N. Hansen. 1954. Variation and speciation in the genus *Fusarium*. Ann. N.Y. Acad. Sci. 60:16–23.
64. Snyder, W. C., and T. A. Toussoun. 1965. Current status of taxonomy in *Fusarium* species and their perfect stages. Phytopathology 55:833–37.
65. Snyder, W. C., H. N. Hansen, and J. W. Oswald. 1957. Cultivars of the fungus, *Fusarium*. J. Madras Univ. B,27:185–92.
66. Tio, M., L. W. Burgess, P. E. Nelson, and T. A. Toussoun. 1977. Techniques for the isolation, culture, and preservation of the Fusaria. Australian Plant Pathol. Soc. Newsletter 6:11–13.
67. Toussoun, T. A., and P. E. Nelson. 1975. Variation and speciation in the Fusaria. Annu. Rev. Phytopathol. 13:71–82.
68. Toussoun, T. A., and P. E. Nelson. 1976. A Pictorial Guide to the Identification of *Fusarium* Species According to the Taxonomic System of Snyder and Hansen, 2nd edition. University Park: Pennsylvania State University Press. 43 pp.
69. Tschanz, A. T., R. K. Horst, and P. E. Nelson. 1975. A substrate for uniform production of perithecia in *Gibberella zeae*. Mycologia 67:1101–8.
70. Ullstrup, A. J. 1936. The occurrence of *Gibberella fujikuroi* var. *subglutinans* in the United States. Phytophatology 26:685–93.
71. von Blittersdorff, R., and J. Kranz. 1976. Vergleichende Untersuchungen an *Fusarium xylarioides* Steyaert (*Gibberella xylarioides* Heim et Saccas), dem Erreger der Tracheomykose des Kafees. Z. Pflanzenkrankheiten Pflanzenschutz 83:529–44.
72. Waite, B. H., and R. H. Stover. 1960. Studies on Fusarium wilt of bananas. VI. Variability and the cultivar concept in *Fusarium oxysporum* f. *cubense*. Can. J. Bot. 38:985–94.
73. Wollenweber, H. W. 1913. Studies on the *Fusarium* problem. Phytopathology 3:24–50.
74. Wollenweber, H. W. 1916–1935. Fusaria autographice delineata. Berlin: Selbstverlag. 1200 Tafeln.
75. Wollenweber, H. W. 1943. Fusarium—Monographie. II. Fungi parasitici et saprophytici. Zentralbl. Bakteriol. Parasitenkd. Infektionskr. II 106:104–35, 171–202.
76. Wollenweber, H. W., and O. A. Reinking. 1935. Die Fusarien, ihre Beschreibung, Schadwirkung und Kekämpfung. Berlin: Paul Parey. 355 pp.
77. Zachariah, A. T., H. N. Hansen, and W. C. Snyder. 1956. The influence of environmental factors on cultural characters of *Fusarium* species. Mycologia 48:459–67.

Index

Page numbers printed in boldface type indicate the principal discussion of a species.